INTEGRATING STEM TEACHING AND LEARNING INTO THE K–2 CLASSROOM

INTEGRATING STEM TEACHING AND LEARNING INTO THE K–2 CLASSROOM

Jo Anne Vasquez
Michael Comer
Jen Gutierrez

Arlington, Virginia

Claire Reinburg, Director
Rachel Ledbetter, Managing Editor
Jennifer Merrill, Associate Editor
Andrea Silen, Associate Editor
Donna Yudkin, Book Acquisitions Manager

ART AND DESIGN
Will Thomas Jr., Director
Jae Martin, cover design

PRINTING AND PRODUCTION
Catherine Lorrain, Director

NATIONAL SCIENCE TEACHING ASSOCIATION
David L. Evans, Executive Director

1840 Wilson Blvd., Arlington, VA 22201
www.nsta.org/store
For customer service inquiries, please call 800-277-5300.

23 22 21 20 4 3 2 1

NSTA is committed to publishing material that promotes the best in inquiry-based science education. However, conditions of actual use may vary, and the safety procedures and practices described in this book are intended to serve only as a guide. Additional precautionary measures may be required. NSTA and the authors do not warrant or represent that the procedures and practices in this book meet any safety code or standard of federal, state, or local regulations. NSTA and the authors disclaim any liability for personal injury or damage to property arising out of or relating to the use of this book, including any of the recommendations, instructions, or materials contained therein.

Library of Congress Cataloging-in-Publication Data
Names: Vasquez, Jo Anne, 1943- author. | Comer, Michael W., 1956- author. | Gutierrez, Jen, 1963- author.
Title: Integrating STEM teaching and learning into the K-2 classroom / Jo Anne Vasquez, Michael Comer, Jen Gutierrez.
Description: Arlington, VA : NSTA Press, [2020] | Includes bibliographical references and index.
Identifiers: LCCN 2019043208 (print) | LCCN 2019043209 (ebook) | ISBN 9781681406206 (paperback) | ISBN 9781681406213 (pdf)
Subjects: LCSH: Science--Study and teaching (Elementary)--United States. | Technology--Study and teaching (Elementary)--United States. | Engineering--Study and teaching (Elementary)--United States. | Mathematics--Study and teaching (Elementary)--United States.
Classification: LCC LB1585.3 .V375 2020 (print) | LCC LB1585.3 (ebook) | DDC 372.35/044--dc23
LC record available at *https://lccn.loc.gov/2019043208*
LC ebook record available at *https://lccn.loc.gov/2019043209*e-ISBN: 978-1-68140-620-6

CONTENTS

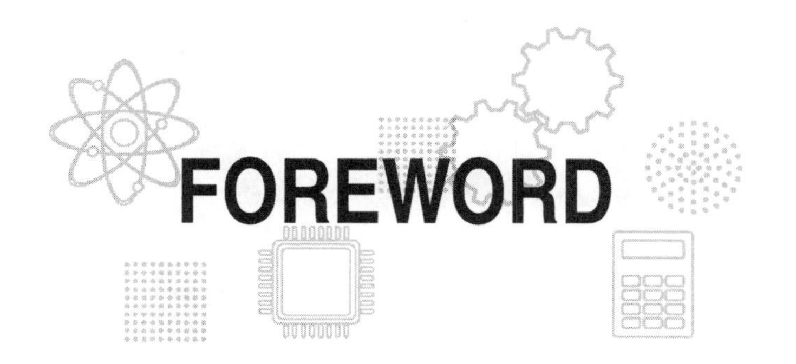

FOREWORD

Integrating STEM Teaching and Learning Into the K–2 Classroom is a critically important contribution toward advancing STEM (science, technology, engineering, and mathematics) education for two overarching reasons: (1) It blazes a trail for early elementary classroom practitioners to reflect the latest thinking in STEM, and (2) it provides a means by which early elementary educators can meaningfully contribute to America's STEM movement.

The course for STEM education across the United States has been mapped in the report *Charting a Course for Success: America's Strategy for STEM Education*, released by the White House Office of Science and Technology Policy in December 2018. Readers and users of *Integrating STEM Teaching and Learning* by Jo Anne Vasquez, Michael Comer, and Jen Gutierrez can be confident of a close alignment between the broad consensus of the STEM education community as reflected in *Charting a Course* and the research-based insights and practical examples provided throughout this book. Common threads woven through both publications are the integration of STEM concepts and principles, the application of classroom experiences to students' lives, a priority on equal access to high-quality STEM education for all learners, the development of interpersonal skills including communication and perseverance, assessment to continuously improve outcomes, and other hallmarks of STEM education.

Today, STEM education has essentially eclipsed its own acronym to be an educational sea change from disciplinary silos toward solving transdisciplinary big questions and problems that converge multiple disciplines, making the enterprise far more interesting to learners. And it starts early, as Vasquez, Comer, and Gutierrez observe—"there is strong evidence that STEM learning can and does begin in early childhood classrooms" (p. 7). Thus, *Integrating STEM Teaching and Learning Into the K–2 Classroom* provides a superb roadmap.

–Jeff Weld
Executive director, Iowa Governor's STEM Advisory Council;
former senior policy advisor on STEM education,
White House Office of Science and Technology Policy

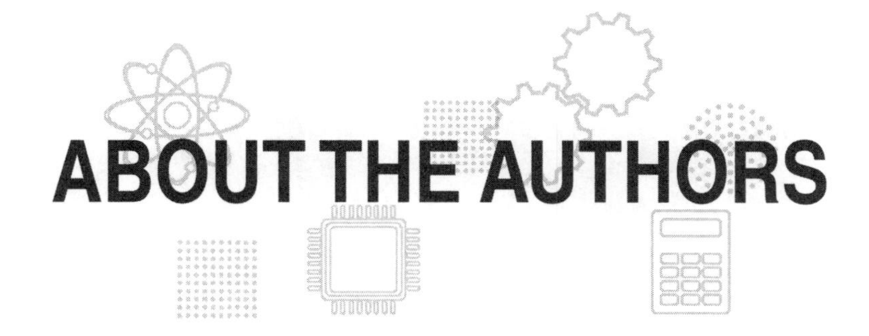

ABOUT THE AUTHORS

Jo Anne Vasquez is a recognized leader in science education. She is a past president of the National Science Teaching Association (NSTA) and the National Science Education Leadership Association and was a Presidential Appointee to the National Science Board, the governing board of the National Science Foundation, becoming the first and only K–12 educator to hold a seat on this prestigious board. She is currently the senior STEM (science, technology, engineering, and mathematics) consultant for Arizona State University's Office of Knowledge Enterprise Development.

Jo Anne's service and contributions to the advancement of science and STEM education at the local, state, and national levels have won her numerous awards: the 2014 National Science Education Leadership Award for Outstanding Leadership in Science Education, the 2013 National Science Board Public Service Award, and the 2006 Robert H. Carlton Award for Leadership in Science Education. She also received the Distinguished Service to Science Education Award, the Search for Excellence in Elementary Science Education and Supervision Award, and the New York Academy of Science's Willard Jacobson Award for major contributions to the field of science and STEM education. In addition, she was the 2004 National Association of Latino Elected and Appointed Officials honoree for her contributions to improving education.

Jo Anne has been involved with curriculum development for McGraw-Hill K–6 science, and she has facilitated STEM education professional learning sessions throughout the United States and in Thailand, China, Singapore, and the Philippines. A graduate of Northern Arizona University, she holds a bachelor of science degree in biology, a master's degree in early childhood education, and a PhD in curriculum and instruction.

Michael Comer began his educational career teaching middle school science in Dobbs Ferry, New York, and Riverside, Rhode Island, before joining the publisher Silver Burdett and Ginn (SBG) as the regional science/mathematics consultant. At SBG, he developed an expertise for providing rich and meaningful workshops that linked the facets of inquiry teaching with hands-on learning using manipulatives. He was an instructor at the Summer Science Seminar at Bridgewater State College in Bridgewater, Massachusetts, for more than 10 years, where he worked with teachers across the K–12 spectrum.

After the merger of SBG with Pearson in 1999, Michael was promoted to science curriculum specialist for STEM products, where he played a vital role in the development of new educational materials to meet changing market demands. In 2004, he joined Macmillan/McGraw-Hill as the national product manager for science and led the product development team in the creation of a brand-new science series, *Science: A Closer Look*, which quickly became a national bestseller. Michael transitioned from marketing to the editorial side of product development in 2013, when he became the editorial director for science and mathematics at Victory Productions in Worcester, Massachusetts, where he directed a team in content development and *Next Generation Science Standards* (*NGSS*) assessment item writing for major assessment providers and educational service organizations.

Michael, who has a BA in biology from American International College, has led many educational workshops on a range of topics in science and mathematics, helping educators embrace the new standards and infuse more hands-on, problem-based learning experiences into their classroom practices. Internationally, he has provided science professional development to teachers in Puerto Rico, St. Maarten, Bahrain, and the Kingdom of Saudi Arabia. His most recent work was to help produce a two-year professional development series for master science teachers in Thailand for the Institute for the Promotion of Science and Technology.

Jen Gutierrez began her educational career in Arizona in 1988, teaching first through fourth grades as well as K–2 multi-age classes. In 2006 she moved into the role of science curriculum specialist at the district level, and in 2014 she joined the Arizona Department of Education in the K–12 Standards Division as the K–12 STEM education specialist. Today she works as a STEM education consultant developing and delivering professional learning opportunities to support educators. She is a proud member of the *NGSS* writing team, including the Diversity & Equity team. She currently serves on the NSTA Board as division director of professional learning.

Jen is interested in three-dimensional teaching and learning, diversity and equity, and science and literacy, which are additional areas that she focuses on in her professional learning work. She did the keynote presentation, "Introduction to the *NGSS*," at the Shanghai International Forum on Science Literacy for Adolescents.

A graduate of Northern Arizona University, Jen holds a bachelor of science degree in journalism, a post-degree certification in elementary education, and an educational leadership–principal certification. She also has a master's degree in elementary education from Arizona State University.

ACKNOWLEDGMENTS

We, the authors of *Integrating STEM Teaching and Learning Into the K–2 Classroom*, would like to thank NSTA Press for supporting this body of work. A special thanks to the editors, especially Jennifer Merrill, our primary editor, for her due diligence in finishing up the manuscript and getting it ready for publication.

We would also like to thank the teachers who were willing to let us come into their classrooms and who shared their exemplary STEM lessons with us:

- Wendy Tucker, W. F. Killip Elementary School, Flagstaff, Arizona
- Lori Schmidt and Josh Porter, Broadmor Elementary School, Tempe, Arizona
- Joe Guiterrez, principal, Killip Elementary, Flagstaff, Arizona
- Allison Davis, kindergarten teacher, Chandler, Arizona

A special thanks goes to contributing author Joel Villegas for his work on Chapter 9.

These wonderful teachers and administrators truly helped this book come alive by providing real lessons and strategies to help teachers as they begin to implement STEM teaching and learning in their classrooms.

An additional special thanks to the countless K–2 teachers we have had the pleasure of working with over the years, who kept asking us to please write a book to support integrating STEM teaching and learning into the real world of a hectic and exciting elementary school day!

Lastly, we would like to acknowledge Jeff Weld, whose NSTA Press book, *Creating a STEM Culture for Teaching and Learning* (2017), provided some valuable foundational knowledge for our manuscript. Jeff also provided the report *Charting a Course for Success: America's Strategy for STEM Education* (Committee on STEM Education of the National Science and Technology Council, 2018). He is assistant director for STEM education at the U.S. Office of Science and Technology Policy.

INTRODUCTION

It's time to ramp up science, technology, engineering, and mathematics (STEM) in the K–2 classroom, according to the Community for Advancing Discovery Research in Education (CADRE). CADRE is a network for STEM education researchers funded by the National Science Foundation's Discovery Research preK–12 program. This new research suggests that high-quality STEM experiences in preK through grade 3 can offer a "critical foundation for learning about these disciplines in ways that facilitate later learning" (Sarama et al. 2018, p. 1).

In particular are the following benefits of early learning in science and math:

- It leads to social-emotional development and fewer challenging behaviors.
- It supports the development of a mind-set that includes curiosity, communication, persistence, and problem solving, among other habits.
- It contributes to gains in all other subjects by supporting literacy and language development and better reading comprehension and writing skills.
- It includes subjects that can engage students from varying backgrounds, including English language learners.

But delivering high-quality early STEM education requires expertise on the part of the teacher in scaffolding the lessons. Among the recommendations offered by Sarama et al. (2018):

- Encourage children to share and elaborate on their observations and ideas, even if they may be incorrect.
- Suggest additional investigations to test students' ideas.
- Provide all children with equal opportunities to participate in STEM experiences.
- Listen to the students and watch them as they play, explore, talk to one another, and engage in STEM activities, to get a sense of what they understand about STEM concepts.

This research lays the foundation for why high-quality STEM teaching and learning is critical in early childhood education; however, the researchers also point out that the teachers themselves need support as they learn how to facilitate STEM learning in their classrooms. Professional learning experiences are needed to cover how teachers can make connections between STEM topics and the everyday activities they are already doing with their students. STEM teaching and learning does not need to become one more add-on to the K–2 classroom. STEM learning should be a natural extension of what teachers are already teaching. It was with this in mind that we set out to write *Integrating STEM Teaching and Learning Into the K–2 Classroom.* We wanted to focus on how to naturally integrate STEM learning into K–2 classroom experiences.

New science standards across the country have keyed in on the idea that student learning is an integration of three central aspects of instruction: (1) tying together the knowledge of core concepts, (2) building fluency in the practices of scientists and engineers, and (3) fostering the ability to make connections across experiences. These things are key to developing long-term student understanding. The STEM lessons detailed throughout this book provide a relevant foundation to help you begin seeing science instruction through this lens as you expand and enhance your K–2 STEM teaching. Practicing K–2 teachers who have developed these exciting STEM lessons and units will take you on their STEM learning journey as they have implemented these experiences in their classrooms. The lessons are developed so that you can teach them in your classroom.

Before you begin your learning journey, we would like to share the following personal experience that author Jo Anne Vasquez had when she began her teaching career. Her story demonstrates that interdisciplinary teaching is not a new strategy for instruction, but is one that has been employed regularly by teachers looking to inspire their students. Many primary school teachers were providing these types of experiences for their students long before it was called STEM.

IT ALL STARTED WITH A CABBAGE

There was nothing unusual about this second-grade classroom. We had reading groups, math time, writing, science, social studies, music, and art—all the usual types of activities you would find in a K–2 classroom. I was so proud. I had survived my first year of teaching and this was the beginning of my second year. With feet now firmly planted on the ground, I knew the routine, and I smugly thought I had it all down. But little did I know that something was about to change the way I thought about teaching forever.

It was the second month into this new school year when the "Aha!" moment happened for me. I was about to get a lesson from my students on how important it is to make teaching and learning relevant for them. I would learn firsthand that using an integrated, interdisciplinary approach to teaching was the key to all of my students' learning. I was about to find that just delivering the content to my second graders was not enough. For these students to really understand and internalize that information, I would need to give them the ability to make connections to everything they were learning and provide opportunities to apply it in meaningful, personal, and grade-appropriate ways.

The Awakening!

The day started off like any other. Class had begun, and we were having our small reading group time at the back table. The story we were reading was about a farmer who had some rabbits that came into his garden and ate all his vegetables. He had carrots, tomatoes, radishes, and cabbages in his garden, and he was so proud because everything was just about ready to harvest. He was going to sell some of his vegetables and also make a wonderful meal for his family. We were talking about the story, and I was doing the usual types of activities that teachers do—checking for understanding, asking questions, calling on students to read certain passages—when all of a sudden a student said, "But teacher, I'm confused."

INTRODUCTION

"Allison, what are you confused about?" I asked. Allison responded, "I've eaten carrots and tomatoes. I've seen radishes, but I've never seen a cabbage." When I questioned the others in the group, only one of the six had either seen or tasted a cabbage. This question alerted me to what I would later term "empty verbalization"—lots of me describing how a cabbage looked and what kinds of dishes you could make with it. All of which seemed to satisfy them for the time being. But who knew where that cabbage question would lead.

The next day I brought in a cabbage. We cut it apart. The children tasted it, felt it, and talked about how it looked and smelled. Of course they wanted to know how and where cabbages grew. And they asked, "Can we grow some?" Well, to make a very long journey a bit shorter, we decided to plant some seeds to watch them grow. We convinced our custodial staff to build some long trays that would hold soil for our plants. There was plenty of sunlight, as we had windows in the room. This became a true learning journey for all of us. We planned for cabbages, carrots, green beans, radishes, and tomatoes. The students learned to measure the plants, kept science logs to record their plants' growth, and checked out books from the library about the different kinds of plants and how they grew. We looked at the parts of the plants, and what began as a reading-group question became a several-months project.

But this story doesn't end with the garden. Many teachers actually do this type of project, so there is nothing unusual in this example. But what happened next was what today we might label a transdisciplinary experience—or more commonly thought of as problem- or project-based learning. When our garden plants were becoming mature enough to begin to harvest, the children wanted to share their plants with their parents. This became the transdisciplinary experience as our class set off to develop a meal, create a menu, and compose invitations to the parents.

Soon it was decided that the students needed help to cook the meal because they had chosen spaghetti. A few phone calls later, and a couple of eager parents agreed to assist. The student committee also asked our cafeteria staff if they could use the kitchen and would the staff join in to help cook the spaghetti. The students did butter their bread and mix their salad. At our feast, the children set the tables, served the plates of food, and spent time talking about what they had learned while growing their own salad. And best of all, Allison showed and described what she had learned about a cabbage. For if it had not been for Allison's question about the cabbage, this adventure would never have taken place.

Conclusion

We might venture to say that many of you reading this might be thinking that this example is no different from what you already do in your classroom. In many ways, interdisciplinary teaching is what preschool and primary teachers have been doing with their students. Most teachers of young students know instinctively that children learn best when they are active participants in the process. What many of you may have already been doing demonstrates the difference between just providing your students with *academic learning* and shifting the acquisition of knowledge to *intellectual learning*.

The differences are very obvious, according to Lilian Katz, professor emerita of early childhood education at the University of Illinois at Urbana–Champaign. Academic learning "is stuff that is clear like the alphabet, it has no logic, it just has to be memorized ... and does have to be learned eventually," Katz states (see Pica et al. 2012). Intellectual learning "has to do with reasoning, hypothesizing, theorizing, and so forth, and that is the natural way of learning."

This research is not new. It seems that years ago in education we understood that students did learn best and retained the information longer when engaged in active learning in which they were predicting, hypothesizing, reasoning, applying, and describing what they were doing. Although it did not have the formal label of STEM education, interdisciplinary learning has been around for as long as students have been going to school. However, somewhere along the way, this active, intellectual learning became an afterthought. Today, we know that with the label of STEM, it has moved from the background to the foreground, and everyone wants to be on board the "STEM train."

Once, a second grader was asked, "What is mathematics?" and he responded, "It is something we do in the morning at school." We want our students to internalize what they are learning and be able to apply this new learning in many different ways. In other words, we want STEM teaching and learning to be part of the whole life of our students, and not just during school hours. During our journey together in this book looking at STEM teaching and learning in the K–2 classroom, we hope you will develop your own operational definition of what it means to be a STEM teacher. We hope you will learn some new strategies and gain some new ideas that you can use in your STEM lessons. We share what the research is saying about why STEM teaching is important in early childhood, with a focus on the K–2 classroom, and perhaps you will realize that you are already a STEM teacher, even if it all begins with a different kind of "cabbage question."

CHAPTER SUMMARIES

This book contains 10 chapters, with the following titles and summaries:

- **Chapter 1: Creating a Blueprint for Building Your K–2 STEM House.** What is STEM education in the K–2 classroom? How is it different from what primary teachers are already doing? How can you do it without adding more to an already full schedule or day? This chapter addresses these questions.
- **Chapter 2: Pioneering Into STEM Integration.** This chapter describes the different levels of STEM integration through the example of a second-grade STEM unit on early Americans and pioneers with English language arts as the driver.
- **Chapter 3: Unpacking the Integrated STEM Classroom.** In this chapter, we identify key elements found in an integrated STEM classroom, detail how they work together, and explore why these elements are critical to a successful student STEM learning experience.

- **Chapter 4: Tackling the Core Instructional Time.** This chapter takes into consideration how the STEM classroom can be used in concert with the core reading block to achieve the goals and objectives essential to all disciplines.
- **Chapter 5: Using the W.H.E.R.E. Model Template.** This chapter introduces the research-based W.H.E.R.E. model template. This template presents a clear and actionable process for curriculum developers and classroom teachers to follow as they develop their own 21st-century STEM experiences.
- **Chapter 6: Developing a STEM Unit With Math as the Driver—Straw Bridges.** The kindergarten STEM unit in this chapter describes how two partnering teachers created an integrated STEM unit using mathematics concepts as the primary instructional focus with the help of fifth graders in the school.
- **Chapter 7: Developing a STEM Unit With Engineering as the Driver—Baby Bear's Chair.** This chapter's kindergarten unit provides an example of an integrated STEM unit with an engineering and design concept as the primary instructional focus.
- **Chapter 8: Developing a STEM Unit With Science as the Driver—A Pond Habitat.** The second-grade unit in this chapter demonstrates how collaboration works in developing an integrated STEM unit using the help of community partners who work in STEM fields.
- **Chapter 9: Moving Students From Inquiry to Application—A Shade Structure.** This chapter's first-grade unit demonstrates how to develop a scenario from anchoring phenomena using a STEM unit that has science as the driver.
- **Chapter 10: Transforming Into a Successful STEM School.** This chapter describes how one district created a successful culture of STEM teaching and learning in its school through the commitment of various education partners working together toward a common goal.

REFERENCES

Pica, R., S. Killins, L. G. Katz, and D. J. Stewart. 2012. "What, Teaching STEM in Preschool, Really?" BAM Radio interview, May 25. *www.bamradionetwork.com/track/what-teaching-stem-in-preschool-really.*

Sarama, J., D. Clements, N. Nielsen, M. Blanton, N. Romance, M. Hoover, C. Staudt, A. Baroody, C. McWayne, and C. McCulloch. 2018. *Considerations for STEM education from preK through grade 3.* Waltham, MA: Education Development Center. *http://cadrek12.org/resources/considerations-stem-education-prek-through-grade-3.*

Creating a Blueprint for Building Your K–2 STEM House

STEM content, when delivered in a context that sparks students' interest and imagination in relatable ways, leads students to see themselves as empowered in the STEM community.

—U.S. Department of Education, *STEM 2026: A Vision for Innovation in STEM Education* (2016)

The question that second grader Allison posed in the classroom, which we shared in the Introduction, was exactly the kind of question that young children often ask. At times they are not inhibited by seeming to be different just because they may not know something or have not seen or experienced a certain thing. Allison said, "But teacher, I'm confused. I've eaten carrots and tomatoes. I've seen radishes, but I've never seen a cabbage." Not only was this an "Aha!" moment for author Jo Anne Vasquez, but also it turned the focus of her approach to teaching from a solely single-subject concentration to more of a meaningful, integrated approach.

Our guess is that with the arrival of the *Next Generation Science Standards* (*NGSS*; NGSS Lead States 2013) and their inclusion of engineering design concepts, some teachers reading this book may be unsure now of just how to teach science and what to teach. Add to this the focus on STEM (science, technology, engineering, and mathematics) teaching and learning, and they may very well feel overwhelmed.

Many of you reading this probably are already teaching with some type of STEM-oriented approach because it came naturally to you. We do not want STEM to be one more teaching strategy that you are required to do, because we recognize you have more than enough on your plate right now. Our goal here and throughout this book is to help many of you realize that you may already be a beginning STEM teacher but just haven't thought of yourself that way. The instructional shifts advocated by the new science standards focus on student understanding by weaving together three distinct dimensions of learning: (1) knowledge of the core ideas, (2) application of the practices (skills) of science, and (3) understanding of the connections that link learning across different topics and multiple subjects.

A Framework for K–12 Science Education encourages us to consider learning as "a developmental progression … designed to help children continually build on and revise their knowledge and abilities, starting from their curiosity about what they see around them and their initial conceptions about how the world works" (NRC 2012, pp. 10–11). This vision focuses on fewer core ideas to study, and it builds across the K–12 learning landscape. The goal of science learning is to help create inquisitive and curious children who as adults can apply their understandings of science and engineering to comprehending and addressing the unanticipated problems of the future.

What is exciting is that STEM teaching provides a vehicle that supports the implementation of this new framework for science education. Building integrated, transdisciplinary units of instruction captures the intent and spirit of the *NGSS*. By introducing engineering tasks and real-world problem-solving challenges as part of the core instruction, we foster that connection in the learning experiences of many 5- to 7-year-olds.

Observing young children at play, you can see them naturally engage in the inquiry and design process. When challenged, "Can you build the tallest tower with connecting blocks so it doesn't fall over?" they immediately set off to work it out, building and testing through trial and error. When asked, "How could you partition the playground so different groups of children can do their favorite activities all at the same time?" they begin asking questions such as "What activities will there be?" or thinking "Is it fair if the spaces are different sizes if all of the activities don't need the same amount of space?" This focuses on their defining the problem to be solved. Figuring out how to get a ball out of a tree is an example of a real-world problem that young children would work to solve in their everyday life.

Some of the skills and strategies we present in this book will be those that you have heard of before and in many ways may already be implementing. Some may be new to you. We hope to inspire you and encourage you to revisit how you integrate your classroom experiences so that they can be even more effective and energizing for your students. So let's get started together on laying the foundation for our early childhood STEM journey.

STEM: FROM ACRONYM TO CLASSROOM PRACTICE

In the 1990s, the National Science Foundation introduced the acronym STEM as the agency continued to receive interdisciplinary proposals that overlapped the directorates within the foundation. In 2007, the National Science Board, the governing board of the National Science Foundation, released *National Action Plan for Addressing the Critical Needs of the U.S. Science, Technology, Engineering, and Mathematics Education System*, putting STEM teaching and learning directly in the forefront of our educational system. Government and private funding began to flow toward all types of STEM education programs, and STEM became recognized as a metadiscipline—an integration of formerly separate subjects into a new and coherent field of study.

To provide a very simple and inclusive definition of STEM, we refer to the one from *STEM Lesson Essentials*:

> *STEM is an interdisciplinary approach to learning that removes the traditional barriers separating the four disciplines of science, technology, engineering, and mathematics and integrates them into real-world, rigorous, and relevant learning experiences for students.* (Vasquez, Sneider, and Comer 2013, p. 4)

This definition is widely accepted, and the beauty of it is that you can take this basic definition and adjust it to fit your own classroom or school philosophy. One certainty is that if you are implementing STEM education in your classroom, some parent will come and ask, "What is this STEM stuff my child is learning?" It is good to have an easy "elevator pitch" type of definition that you can use to describe this approach. You can explain to parents in two or three sentences how this interdisciplinary style of teaching will help their children apply what they are learning in the disciplines in new and creative ways through meaningful, relevant experiences that will prepare them for life outside the classroom.

LAYING THE FOUNDATION FOR BUILDING YOUR K–2 STEM HOUSE

Author Jo Anne's cabbage experience (described in the Introduction) began laying the foundation for the development of a fully integrated, interdisciplinary unit of study for her second graders. STEM, of course, was not a coined term back then. What was coming to the forefront in education research was the idea of integration, which was loosely defined as keeping concepts alive through the practice of having students apply them (Harlan and Rivkin 2012). Integrated methods were not as direct as the teaching of the discipline and depended more on the ability of teachers to recognize how they could keep the new concepts alive by capturing the opportune moment to create projects where the students would apply them.

Building the STEM House

When we think of what STEM would look like in the classroom, we look to the following descriptors for each of the component disciplines:

- *Science is a way of thinking.* Science is observing and experimenting, making predictions, sharing discoveries, asking questions, and wondering how things work. If the students are asking questions and trying to find the answers, it is science.
- *Technology is a way of doing.* Technology is using tools, being inventive, identifying problems, and making things work. It can be an object, a system, or a process. Technology is not just a device or a "plug-in." It is any type of tool that humans have developed to make their lives or work easier.
- *Engineering is the process of solving problems.* Whether using a variety of materials, creating designs, building prototypes, or improving existing solutions, if students are trying to solve a problem, they are engaged in the process of engineering.

- *Mathematics is the process of applying.* Doing operations with numbers provides a way of measuring, sequencing, patterning, and exploring shapes, volume, and sizes; and it allows students to quantify their observations or solutions.

In a learning environment that is intentionally designed to provide meaningful, relevant, real-world experiences for all children, students will very quickly begin to make connections among all of their learning experiences and not see school as just isolated blocks of instructions. According to a *Hechinger Report* article,

> *It turns out that explicitly teaching students about the connections between engineering, science and math, teaching the engineering design process rather than just posing an engineering challenge, and helping students gather information from failed attempts all make a difference to students' ability to absorb and retain science and engineering concepts.* (Mongeau 2019, p. 5)

The research is beginning to reveal that when elementary schools introduce young learners to the principles of engineering early, they inspire a lifelong interest in STEM-related fields. A review of a variety of reports, presented papers, and research studies by the National Institute for Early Education Research at Rutgers Graduate School of Education, the National Association for the Education of Young Children, and the American Educational Research Association shows the promise of how rich science and math experiences help foster language and literacy skills in students, which is carried forth as they progress in school.

Reading and Writing to Learn!

In the integrated cabbage unit, the learning took on a life of its own. The second-grade students made observations and drawings of the seeds before they were planted. There was a discussion about what the seeds would need to grow. The students realized they would need to provide them with light, soil, and water. Students were assigned to groups, and they were to take turns watering the plants. A class chart was created to provide a record of the watering schedule and when it was completed. All students were responsible for keeping their own plants log and were tasked with measuring and recording their plants' growth. For some students, the process of just being organized and keeping to a schedule was a real awakening for them. Also, it was amazing how quickly all of the students began to take ownership of the window gardens. These were their plants!

Comparing the size and variety of leaves led to a class discussion about why some plants had very fine leaves and others had darker, thicker leaves. It was not critical at this point for the students to know or even understand why there were these differences. What was important was the fact that they were using their observation skills, and most of all they were describing and discussing what they were seeing. It also became evident that all of a sudden students were aware of leaves everywhere. In the school yard, in their neighborhoods, and even in their own yards and homes, leaves were discovered and discussed. They began to inquire, "What is the function of the leaves and stems?" Lots of intellectual learning, but the more important aspect was that they were having an awakening.

As *A Framework for K–12 Science Education* (NRC 2012) outlines, the development of the core ideas reflects a progression of understanding that builds with repeated exposure to those core ideas over time. For example, if taught today, the life science (LS) core idea in the cabbage unit might focus student attention on several aspects of LS1 (From Molecules to Organisms: Structure and Function), where the overarching question is "How do organisms live, grow, respond to their environment, and reproduce?" Concentration on what students show understanding on is provided in the endpoint descriptions of the grade bands. For K–2, the various student understandings for three different aspects of the LS1 disciplinary core idea are as follows:

- "*LS1.A: Structure and Function:* All organisms have external parts. … Plants also have different parts (roots, stems, leaves, flowers, fruits) that help them survive, grow, and produce more plants" (NRC 2012, p. 144).
- "*LS1.B: Growth and Development of Organisms:* Plants and animals have predictable characteristics at different stages of development. Plants and animals grow and change. Adult plants and animals can have young" (NRC 2012, p. 146).
- "*LS1.C: Organization for Matter and Energy Flow in Organisms:* Plants need water and light to live and grow" (NRC 2012, p. 147).

The Cabbage Unit as a Well-Designed STEM Lesson

The problem scenario: We have planted our garden and taken care of our garden, and now what are we going to do with the plants from our garden? Not knowing then that a STEM transdisciplinary activity should begin with a problem scenario to hold students' interest as they navigate through the STEM learning, it just came about naturally as the students brainstormed ways to share their garden. The students' spaghetti dinner with their homegrown salad became the culmination of what they had begun working toward when they started their garden project.

In reflecting back on this unit, we see that it had science, mathematics, technology, engineering design, and even aspects of English language arts (ELA) woven into its activities. Beginning with the planting itself to the way the students decorated the tables, designed and created signs and menus, and developed their presentations to describe all they had learned about their plants, their learning was tightly connected starting with the initial question.

These students not only participated in a STEM unit, they also used their ELA skills with their writings and communications. Most important, the students had a reason to learn, and what they were learning they were applying. STEM education is about having students apply what they have learned in the disciplines in meaningful ways. It's all about application, application, application.

CHAPTER 1

Nurturing STEM Skills in the K–2 Classroom

To support the future of our nation, the seeds of STEM need to be planted early, along with, and in support of, the seeds of mathematical and reading literacy. Together, these mutually enhancing, interwoven strands of learning will grow well-informed, critically thoughtful citizens prepared for the challenges of tomorrow.

Why then is STEM learning not woven more seamlessly into early childhood education? An examination of the environments and systems in which children live reveals that it is not due to a lack of interest or enthusiasm on the part of the students, teachers, or parents. The barriers are more complex. In December 2013, the National Science Foundation, the Smithsonian Institution, and the Education Development Center cohosted a STEM Smart workshop to ready early childhood practitioners. Participants were delighted to learn of evidence-based practices and tools, but many declared that they felt too constrained by the current school structures and policies to apply these findings. They also found that the public holds several misconceptions about STEM learning. These beliefs include the notion that STEM is primarily for older students and that children in the lower grades need to learn basics first, that STEM is only important for those who especially excel in these areas or look at them as career opportunities, or that STEM and other learning topics must be taught separately from the existing curriculum. When advocates and communicators of STEM teaching and learning do not carefully frame their messages, they can inadvertently activate and strengthen these misconceptions.

Several other recommendations also were proposed at the workshop, such as engaging parents who can be their children's first STEM guides, supporting teachers through improved professional learning, and offering greater support for teaching early childhood STEM. The early research around nurturing STEM skills in the K–3 classrooms provides support for including STEM in the early childhood classroom. As noted in *A Framework for K–12 Science Education*, "before they even enter school, children have developed their own ideas about the physical, biological, and social worlds and how they work. By listening to and taking these ideas seriously, educators can build on what children already know and can do" (NRC 2012, p. 24). Yet current data on school readiness and early mathematics and science achievement (data on the *T* and *E* of early STEM learning are not available) indicate that we are not giving young children the support they need to be STEM smart.

All approaches to nurturing preK through third-grade children's STEM skills and knowledge should reflect the following eight indicators of an effective K–3 curriculum, as identified by the National Association for the Education of Young Children and the National Association of Early Childhood Specialists in State Departments of Education (2009):

1. Children are active and engaged.
2. The goals are clear and shared by all.
3. The curriculum is evidence based.

4. Valued content is learned through investigation, play, and focused intentional teaching.
5. The curriculum builds on prior learning and experiences.
6. The curriculum is comprehensive.
7. Professional standards validate the curriculum's subject-matter content.
8. Research and other evidence indicate that the curriculum, if implemented as intended, will likely have beneficial effects.

Ensuring that every child has a high-quality early STEM education is one of the best investments our country can make. Tomorrow's engineers are building bridges in their classroom's block corner today. Tomorrow's scientists are doing "field work" at recess, inspecting the structure of a fallen leaf or the mechanisms of how an inchworm moves. We have to keep these students questioning, exploring, investigating, and wondering as we work to ensure they have strong K–3 STEM learning experiences. Cultivating curiosity can foster the habits of inquisitiveness for a lifetime.

Jeff Weld, in his book *Creating a STEM Culture for Teaching and Learning*, states the following:

> ***Why STEM Matters to a Teacher.*** *The proverbial rubber meets the road when teachers match-make students and STEM in the classroom. If reading and writing were once the great equalizers that provided a passport to a good living for any American willing to work at it, today it is STEM skills such as coding, statistics, logic, reasoning, critical thinking, creativity, collaboration, and problem solving.* (Weld 2017, p. 16)

Yes, there is strong evidence that STEM learning can and does begin in early childhood classrooms. Some of the teaching strategies and units you have already created are probably much more STEM oriented than you might think. In other words, you may already be a STEM teacher! Our goal with this book is to take you on a learning journey; for some of you, it will be new information, and for others it will be a validation of the teaching strategies you are already using. We hope that all of you will gain some new insights and ideas to use in your classrooms.

Safety Considerations for the Activities in This Book

STEM teaching necessarily involves working with different materials, and at times this can pose safety hazards. Safety always needs to be the first concern in all our teaching. Teachers need to ensure that their rooms, playgrounds, and other learning spaces are appropriate for the activities being conducted. That means that personal protective equipment (PPE) such as sanitized safety glasses with side shields or safety goggles are available and used properly. PPE is to be worn as appropriate during all components of investigations (i.e., the setup, hands-on investigation, and cleanup) when students are using potentially harmful supplies and equipment. At a minimum, the eye protection PPE provided for students to use must meet the ANSI/ISEA Z87 standard. Remember to also review and comply with all safety policies and procedures that have been established by your place of employment. Teachers also must practice proper disposal of materials.

The National Science Teaching Association maintains a website (*www.nsta.org/safety*) that provides guidance for teachers at all levels. The site also has a safety acknowledgement form (*http://static.nsta.org/pdfs/SafetyAcknowledgmentForm-ElementarySchool.pdf*), which is sometimes called a "safety contract," specifically for elementary students to review with their teachers and have signed by parents or guardians. It cannot be overstated that safety is the single most important part of any lesson. Safety notes are included in this book to highlight specific concerns that might be associated with a particular lesson.

The safety precautions associated with each investigation are based, in part, on the use of the recommended materials and instructions, legal safety standards, and better professional safety practices. Selecting alternative materials or procedures for these investigations may jeopardize the level of safety and therefore is at the user's own risk. Remember that an investigation includes three parts: (1) the setup, which is what you do to prepare the materials for students to use; (2) the actual investigation, which involves students using the materials and equipment; and (3) the cleanup, which includes cleaning the materials and putting them away for later use. The safety procedures and PPE we stipulate for each investigation apply to all three parts.

REFERENCES

Harlan, J. D., and M. S. Rivkin. 2012. *Science experiences for the early childhood years: An integrated affective approach.* Boston: Pearson.

Mongeau, L. 2019. How to build an engineer: Start young. *The Hechinger Report*, January 24. *https://hechingerreport.org/how-to-build-an-engineer-start-young.*

National Association for the Education of Young Children and National Association of Early Childhood Specialists in State Departments of Education. 2009. Where we stand on curriculum, assessment, and program evaluation. *www.naeyc.org/files/naeyc/file/positions/pscape.*

National Research Council (NRC). 2012. *A framework for K–12 science education: Practices, crosscutting concepts, and core ideas.* Washington, DC: National Academies Press.

National Science Board. 2007. *National action plan for addressing the critical needs of the U.S. science, technology, engineering, and mathematics education system.* Washington, DC: National Science Foundation.

NGSS Lead States. 2013. *Next Generation Science Standards: For states, by states.* Washington, DC: National Academies Press. *www.nextgenscience.org/next-generation-science-standards.*

U.S. Department of Education, Office of Innovation and Improvement. 2016. *STEM 2026: A vision for innovation in STEM education.* Washington, DC: U.S. Department of Education.

Vasquez, J. A., C. Sneider, and M. Comer. 2013. *STEM lesson essentials, grades 3–8: Integrating science, technology, engineering, and mathematics.* Portsmouth, NH: Heinemann Press.

Weld, J. 2017. *Creating a STEM culture for teaching and learning.* Arlington, VA: NSTA Press.

Pioneering Into STEM Integration

STEM is an interdisciplinary approach to learning that removes the traditional barriers separating the four disciplines of science, technology, engineering, and mathematics and integrates them into real-world, rigorous, and relevant learning experiences for students.

—J. A. Vasquez, C. Sneider, and M. Comer, *STEM Lesson Essentials* (2013)

STEM (science, technology, engineering, and mathematics) teaching and learning represents a fundamentally different approach to organizing your classroom curriculum. As such, it raises a number of practical questions. What does *integration* really mean? Is it sufficient for students to see the connections between concepts in different fields? Or does it mean assimilating concepts from two different fields so they become one? Should integration involve skills as well as concepts? How about connections outside of the core STEM areas to other disciplines such as social studies or English language arts (ELA)? Or to everyday life?

WHAT INTEGRATION REALLY IS

A Framework for K–12 Science Education (NRC 2012) provides a vision of science instruction that proposes that students' learning experiences engage them in questions about their world, build understanding over time, and prepare them with the skills and curiosity to appreciate the importance of science and engineering as a human endeavor. The goal is to create a more coherent approach to science education in which learning is viewed as a developmental progression of core ideas with increasing levels of sophistication, such that the capabilities and abilities of critical thought result in the learner having a greater capacity for understanding and dealing with the issues facing a future society.

According to the *Framework*, "science is not just a body of knowledge that reflects current understanding of the world; it is also a set of practices used to establish, extend, and refine that knowledge" (NRC 2012, p. 27). The vision of the *Framework* is the weaving together of three dimensions of learning: content, practices, and crosscutting concepts. How these dimensions get integrated into practical learning experiences is left to the classroom practitioner.

CHAPTER 2

There are multiple models for developing integrated units for classroom instruction. Each model proposes different levels of integration that reflect the goals of that particular instruction. Today's instruction shifts the emphasis of teaching from a checklist of "what to teach" to more of a menu of experiences that fosters student understanding and asks them to demonstrate "what they know." Like the *Next Generation Science Standards* (NGSS Lead States 2013), the emphasis of learning is placed on expected student outcomes from what was taught and not on disconnected chunks of information that were presented.

And recent research supports the focus on an integrated approach to learning. As reported in *STEM Integration in K–12 Education*, "The findings suggest that integration can lead to improved conceptual learning in the disciplines but that the effects differ, depending on the nature of the integration, the outcomes measured, and the students' prior knowledge and experience" (NRC 2014, p. 52). The report does caution,

> *It is not surprising that very little is known about how to organize curriculum and instruction so that emerging knowledge in different disciplines will mesh smoothly and at the right time to yield the kind of integration that supports coherent learning. Without very careful attention to developing coherent knowledge structures, the danger is that one or more of the "integrated" disciplines will receive short shrift in its development.* (NRC 2014, p. 53)

Therefore, it is important to "highlight the need to carefully frame the instructional goals and settings to support students in making links to concepts in science" (NRC 2014, p. 52).

Before we jump into creating integrated STEM learning experiences, let's first examine some different levels of integration as a basis for implementing more meaningful STEM experiences in the classroom.

THREE APPROACHES

STEM integration is not one thing. Many educators, when they talk about STEM, describe an approach similar to problem-based or project-based learning (PBL). True, this is one level of STEM integration, but there can be other levels of integration as well. In this chapter we describe what we mean by the different levels of STEM integration through a second-grade "Early America and Pioneers" unit. You might ask yourself, "What, a unit on pioneers? What does that have to do with STEM teaching and learning?" Well, let's find out!

We would like to introduce you to Wendy Tucker, a second-grade teacher at W. F. Killip Elementary School in Flagstaff, Arizona. You will read more about the Killip school. Over the past four years, the teachers at Killip have been developing STEM units for their grade levels. They have received extensive STEM professional learning and have a designated STEM teacher to help lead them. The team of teachers from each grade level works together to develop the STEM units. This is exactly what Wendy's team accomplished by developing a pioneer unit.

Wendy has been teaching for 30 years and has spent the last 20 of them at Killip. Observing in Wendy's classroom, you come to realize that all the tenets of what educators know about teaching primary school students are in place. The students are given opportunities to collaborate and to work independently. Most of all, they are provided with many opportunities to share their ideas and experiment with different approaches, which you will see as the pioneer STEM unit unfolds.

STEM Integration as a Continuum

Integrating STEM teaching and learning should not be thought of as the final product of a single type of learning experience. However, it can be helpful to think of integration as a series of gradually more comprehensive and interwoven learning experiences moving toward a fully integrated PBL-type unit. To illustrate this progression, we've positioned different models of integrated learning on an inclined plane. This progression of models begins with *disciplinary learning* and moves toward integration through *multidisciplinary* or *thematic integration*, *interdisciplinary integration*, and finally *transdisciplinary integration* (see Figure 2.1). Teaching in a STEM-integrated way does not mean abandoning the content standards. The standards are the anchor for all STEM units. The disciplinary core ideas and the science and engineering practices in the individual science activities or investigations are the foundation for the student experiences. How the students engage in that experience and begin to make sense of it in a much larger context is where the crosscutting concepts come into play.

Figure 2.1. Levels of STEM integration

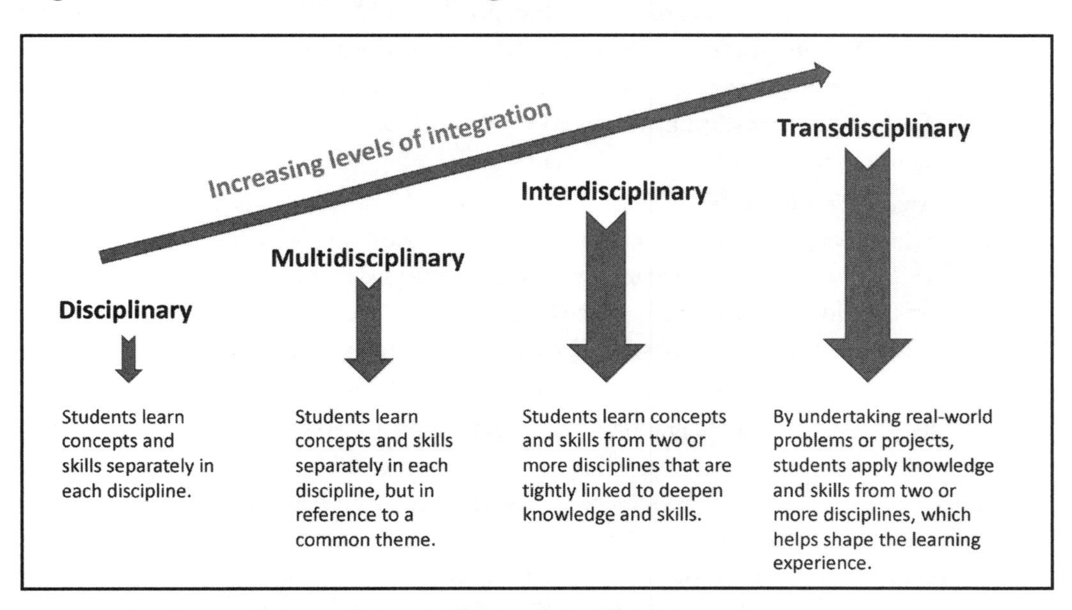

Source: Vasquez, Sneider, and Comer (2013, p. 73). Reprinted with permission.

At the bottom of this continuum plane sits disciplinary teaching, where the students are learning the content and skills of the single subject. The highest point of the continuum is transdisciplinary integration, which is where the fully integrated PBL experience comes in. As we move up the plane there are two other approaches to organizing the STEM curriculum in an integrated way.

Multidisciplinary integration, or the thematically integrated approach, means that the teaching of the concepts and skills found in the separate subjects is linked through a common theme. Multidisciplinary integration is very common in many primary classrooms. For example, as you think about thematic integration, the name of this unit conjures up how the student learning experiences may be linked together. A simple study of the early American settlers can include content from the areas of reading, science, mathematics, and, of course, social studies. The pioneers theme connects the learning experiences in the Early America and Pioneers STEM unit and is relatively easy to design. The students may read about different pioneers and their hardships. They may study landforms and maps and look at the different routes that pioneers took to reach their destinations. They may solve mathematical problems using pioneers as context for their word problems. These individual learnings tie back to the pioneers theme, but the explicit learning is still grounded in the different individual disciplines. This is one type of integration that you might have been using already and not realized that it could be a level of STEM integration.

However, let's look at integration more deeply. Moving from multidisciplinary we shift to interdisciplinary, where the curriculum unit is organized around common learning concepts that span disciplines. Here, the disciplinary concepts and skills become interconnected and interdependent, and the lines between those disciplines become more blurred. The broader crosscutting concepts serve as the focal point of the integration. For instance, in this pioneer unit, the students are not just performing random measuring activities in the context of learning about pioneers, they are performing explicit measurements and data collection as part of developing an understanding of cause and effect. They are measuring how changes to their wheels and axles will affect how far their covered wagons can travel and stay on their "trail."

The students will apply their math skills as they gather evidence as well as demonstrate their reasoning in their observation and communication skills. They look at how landforms influenced the pioneers' travel. They ask questions about why the pioneers needed to travel away from their homes. In other words, the focus on learning is on the bigger crosscutting concept of cause and effect. The teaching of the individual disciplines is not separated as science time, math time, and ELA time but is experienced by the students as the natural built-in learning in the unit as a whole.

The highest level of STEM integration is the transdisciplinary level, which is, as stated earlier, sometimes referred to as PBL. As you will see in the following description of the Early America and Pioneers STEM unit, the transdisciplinary experience is established from the beginning with the presentation of a problem scenario that sets up the overall learning goals and provides students with a direction for learning. Through this model, the students will be engaged in active problem solving, exploring key disciplinary core ideas while using a variety of skills as they work to solve the initial challenge problem. The unit becomes much more meaningful to them as they have a purpose in what they explore and experience.

Wendy Tucker and her grade-level team developed the pioneer STEM unit by applying their standards, the resources they had available, and use of the levels of STEM integration. The following provides a comprehensive overview of how they moved from the standards they were already teaching, along with literature selections to help set the context of the pioneers in American history, and provided opportunities for students to build their own wagons and develop an understanding of the application of measurement.

Curious now? Let's more closely examine how this unit could unfold if we consider it through the lens of the individual levels of integration. Compare how the different levels of STEM integration influence the work that the students would do through this one example, the second-grade Early America and Pioneers STEM unit.

Disciplinary Teaching

At the beginning of the levels of integration plane sits disciplinary teaching, in which students learn the content and skills of individual subject areas separately. Here the presentation and assimilation of subject-specific content is the ultimate goal. Accumulation of information often drives the learning experience, without much emphasis on how or where to use that information. If this unit was strictly disciplinary, the focus would be on just the core learning targets in reading, which in this case emphasize the second-grade reading standards (e.g., RI2.1) and the district learning targets (e.g., all student readings). In the Early America and Pioneers STEM unit, the team began with reading, in which Arizona ELA standards (based on the *Common Core State Standards*) call for the following:

- **RI2.1.** Ask and answer such questions as *who, what, where, when, why,* and *how* to demonstrate understanding of key details in a text. (All student readings)
- **RI2.3.** Describe the connection between a series of historical events, scientific ideas or concepts, or steps in technical procedures in a text. (Pilgrims, colonial days, pioneers)
- **RL2.1.** Ask and answer such questions as *who, what, where, when, why,* and *how* to demonstrate understanding of key details in a text. (All)
- **RL2.2.** Recount stories, including fables and folktales from diverse cultures, and determine their central message, lesson, or moral. (Folktale unit)
- **RL2.3.** Describe how characters in a story respond to major events and challenges. (All)
- **RL2.5.** Describe the overall structure of a story, including describing how the beginning introduces the story and the ending concludes the action. (All)

These standards form the foundation for the district learning targets.

The second-grade team worked together to create a set of specific and measurable learning targets for their students. These learning targets are presented as "I can …" statements that display student outcomes as a series of positive acclimations:

- I can answer questions such as *who, what, where, when, why,* and *how* to show understanding of a story.
- I can read and understand informational texts about history.
- I can tell how events in history are connected.
- I can understand ideas in nonfiction.
- I can retell a story.
- I can tell how characters in a story respond to major events and challenges.
- I can find and understand the beginning, middle, and end of a story.

These skills fit right in with the literature selections the team had designated to use for the Early America and Pioneers unit, which are *The Courage of Sarah Noble* by Alice Dalgliesh and *Wagon Wheels* by Barbara Brenner.

In the disciplinary approach, the identification of standards, outcomes, and selected resources would be enough to teach this unit. Getting the students excited about the early explorers and the pioneers through these literature selections becomes a different challenge. How to make this learning engaging, memorable, and applicable helped set the stage for transitioning it from a disciplinary approach to a more integrated STEM learning experience. This is why interdisciplinary experiences help students access the content better, because they actually apply it in other areas. It is through an exciting STEM unit that the content can come alive for the students.

Multidisciplinary Integration

Probably if we had to pick one level of integration for the Early America and Pioneers STEM unit to fit into, it would be the multidisciplinary level. There is a unifying theme that connects the different content areas, and the content-specific knowledge and skills are still set up as learning goals. Many K–2 teachers teach in this way and have a "theme" as their focus. The challenge here is to start at this level of integration and then move it up the STEM integration plane so that it becomes transdisciplinary.

If we examine this STEM unit through the lens of multidisciplinary integration, you will notice that the unifying theme of early America and pioneers connects the different content areas. The acquisition of content-specific knowledge and skills is still the learning goal, but there is an attempt to loosely link the learning together using broad-based attributes or general commonalities. Multidisciplinary integration has advantages, particularly when the second-grade team of teachers at Killip Elementary School began their initial planning of this STEM unit. There is an explicit intent within the instruction to connect the knowledge, skills, and understandings together in a way that will engage and make sense to the students. These types of integrated learning experiences may

Figure 2.2. Multidisciplinary integration approach for the Pioneers STEM unit

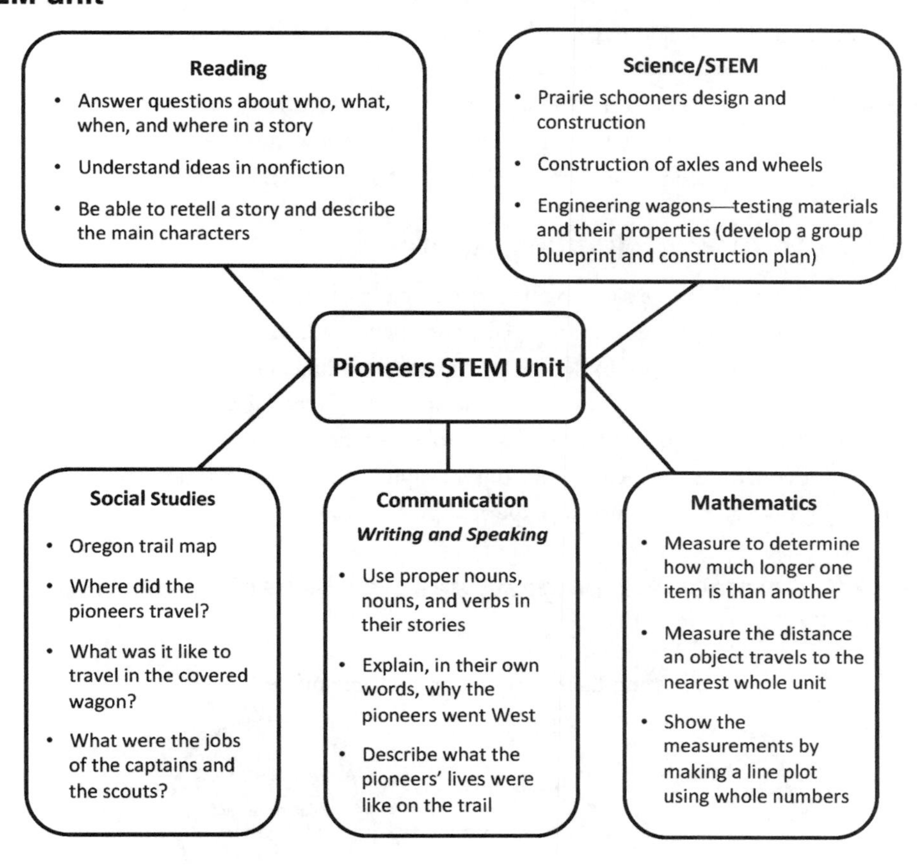

also eventually lead to facilitating an increased level of integration as a natural outgrowth of initial shared planning—which is exactly where this team took the learning (see Figure 2.2).

The diagram identifies the content connections and shows how Wendy and her team began the planning and development of this STEM unit. Reading standards were still the core learning objective. In science, while their engineering task around the concepts of force and motion became the construction of a covered wagon to tie to the pioneer theme, the instruction still focused on the individual science standards. In math, with the work on computation skills along with the concept of greater than and less than, and practice with measuring, the unit used pioneer-related examples as a way to tie to the theme. Writing time focused on creating sentences about pioneers and allowed for developing practice in identifying nouns and verbs. In social studies, the teaching of map-reading skills used examples of the Oregon Trail. As you can see, the students' connections among the individual subjects were common, yet superficial. Their individual learning outcomes were still tied directly to the goals of the individual grade-level disciplines.

Many primary-level teachers use a theme approach to their teaching. While this has a positive impact on students by making their learning seem "connected," it is important that the integration be purposeful in the application. If there is little or no real connection between the content areas under the umbrella theme, students can become confused about the goals of the learning and bored with the exercises because the connections seem contrived. As the second-grade Killip team continued their discussion of the levels of STEM integration, you will see how they incorporated higher levels of integration as they continued in the further development and refinement of their STEM unit.

Moving to an Interdisciplinary Approach

Using the interdisciplinary approach to integration, the teachers were able to reorganize the curriculum around more common learning goals, and they emphasized the interdisciplinary skills and broader connecting concepts. In the Early America and Pioneers STEM unit, a common crosscutting concept within the disciplines was "cause and effect" (see Figure 2.3). The disciplines are still identifiable but tend to be fused more closely together than in the multidisciplinary approach. The students move with greater ease from one discipline to another as they apply the bridging concepts that link the common learning experiences.

Figure 2.3. Common learning: A cause-and-effect diagram for the Pioneers STEM unit

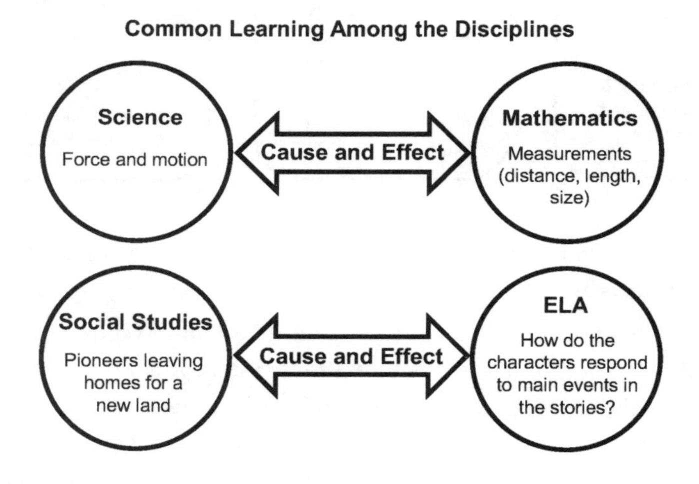

The instructional intent of the unit shifts from a focus on the individual discipline's content targets to a more explicit use of cause and effect as students engage in the unit activities. As the students record measurements of the distance their covered wagon traveled, it leads them to make changes to the size of the wheels and the length of the axles to improve their engineered vehicle. Understanding how changes to the wheels affect the distance the wagon travels requires that they perform even more measurements. The boundary between science and math becomes blurred.

From a humanities perspective, learning the reasons why some people moved West was made more understandable by reading related stories in literature. Integration at the interdisciplinary level is usually tighter and focuses on fewer disciplines to keep the connection stronger.

Moving to Transdisciplinary Integration

The final version of the Early America and Pioneers STEM unit demonstrates the highest level of integration, mainly because of the observations of Wendy and her second-grade team. They recognized places in the unit where the students could take ownership of the learning themselves and apply their developing understanding and newly acquired skills if there was a more relevant task. In this unit, Wendy and her team moved to a transdisciplinary approach by adding an overarching problem scenario to unite the different disciplines together. The students experienced making their wheels and axles, along with creating blueprints for their wagon designs, as part of their tasks (see Figure 2.4).

Figure 2.4. Students working on their wagon designs

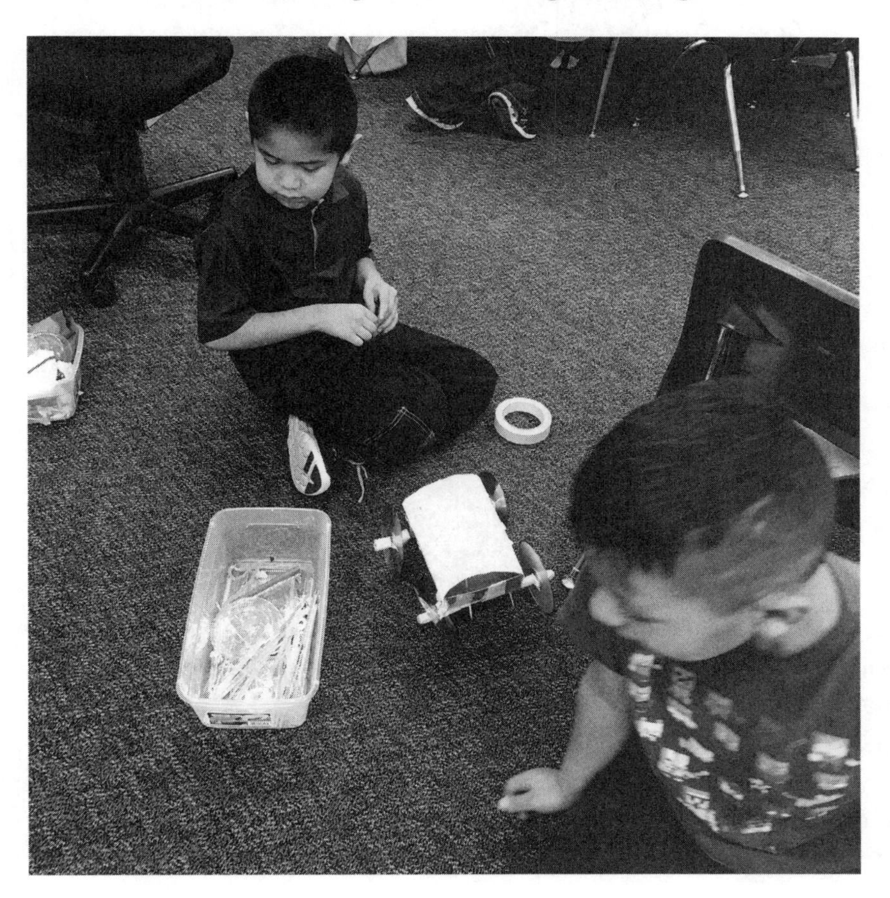

The students had been working on their measurement skills and of course using the interdisciplinary common concept of cause and effect to help them in the construction of their wagons and in their analysis of the literature they were reading. The students were excited to show off their completed covered wagon (see Figure 2.5).

Figure 2.5. Students showing off their completed covered wagon

The real-world application of the students' experience in the Early America and Pioneers STEM unit became the focal point when the goal was to answer the driving question that was posed to them from the context-setting scenario.

The scenario for the transdisciplinary task was as follows: "The pioneers have decided it is time to head out West, and they must decide what they are going to take on their journey and how to construct their wagon so that it will be able to make the long trip."

The student teams needed to answer the following driving question: "Can your team build a wagon with working axles, so that the wheels actually roll, and when tested will travel at least 6 feet rolling down a ramp?"

With their driving question to guide them, the students all became "covered wagon engineers" and were tasked to design, construct, and test their team's wagon. They used what they were reading in their literature selections to understand the requirements for their wagons. They could understand what the "route" was like from social studies and how they needed the wagon to stay on a path. The ramp became their path, and the goal was to keep their wagon from running off the sides of the "trail" (see Figure 2.6). The overarching task focused on the literature and science standards as the drivers for the learning. These standards were supported by the mathematics and social studies standards that fill in around the main task.

Figure 2.6. Students preparing to have their covered wagon roll down the "trail"

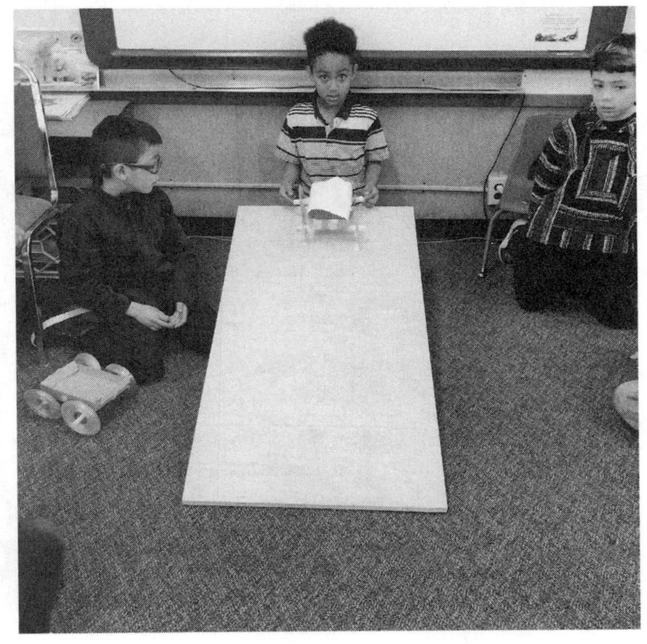

SUMMARY

All of the approaches to integration were woven into the Pioneers STEM unit. Wendy and her second-grade team were able to plan and teach this unit in a way that helped their students see the connections in their learning among all of the different content areas. And the learning was further enriched by integrating authentic assessment tasks, and by using a variety of integrated tasks. There is not one way to approach integrated curriculum development. As shown in this STEM unit, it can be a culmination of all the approaches. The teachers took it to the highest level of integration, which is the transdisciplinary level—always remembering that students can't apply what they don't know. The content is still taught, but what STEM teaching and learning does for students is to take it to the next level. It answers the question "Why do I need to know this?"

REFERENCES

National Research Council (NRC). 2012. *A framework for K–12 science education: Practices, crosscutting concepts, and core ideas.* Washington, DC: National Academies Press.

National Research Council (NRC). 2014. *STEM integration in K–12 education: Status, prospects, and an agenda for research.* Washington, DC: National Academies Press.

NGSS Lead States. 2013. *Next Generation Science Standards: For states, by states.* Washington, DC: National Academies Press. *www.nextgenscience.org/next-generation-science-standards.*

Vasquez, J. A., C. Sneider, and M. Comer. 2013. *STEM lesson essentials, grades 3–8: Integrating science, technology, engineering, and mathematics.* Portsmouth, NH: Heinemann Press.

CHAPTER 3

Unpacking the Integrated STEM Classroom

Society and the demands of the workforce are changing at a rapid rate, as is our perception of what to teach children and what they need to know to survive. The world of children and young people outside of schools has changed, and so the school environment, teaching methods, and the content aren't relatable or inspiring to them any longer, which creates motivational problems.

—Kristiina Kumpulainen, professor of pedagogy, University of Helsinki

We begin this chapter by identifying the key elements found in an integrated STEM (science, technology, engineering, and mathematics) classroom, detailing how they work together, and explaining why these elements are critical to a successful student STEM learning experience.

The most effective, and certainly most exciting, way to engage our students in learning is to make it real and relevant. In other words, make it about them. When students feel connected to the content, behavioral issues tend to take less time away from instruction. Students are too busy exploring the learning, their hands and minds engaged in the activities, to give into distractions that often take them off course and into trouble or timeouts.

It is the essence of STEM teaching and learning that helps us bring the students back into the focus of learning. Regardless of which discipline is leading the charge, students are the driving force behind successful teaching with their curiosity and their questions. Young students are curious by nature, and when we remove the critical obstacles that prevent them from personally connecting to the concepts being taught, we find that they are more likely to to listen, follow along, pay attention, and learn. According to the research reviewed in *STEM Integration in K–12 Education*,

> *Social and cultural factors are fundamental to all learning experiences and particularly important in integrated experiences, which typically require students to work with each other and actively engage in discussion, joint decision making, and collaborative problem solving. Integrated STEM education often involves*

extensive collaboration among teachers and students, and therefore its success depends on the design and effectiveness of the social aspects of the approach. (NRC 2014, p. 86)

When you begin to think about what a STEM teaching and learning environment looks like, sounds like, and feels like, you may recognize many features that you are currently employing in your classroom. There are myriad research articles, documents, and books available that describe a STEM learning environment, including the physical layout, the challenges of the cognitive tasks, and the sociocultural structures of the environment. Learning, especially in the classroom setting, is supported by many different types of social interactions—observations and imitations of peers; the ability to focus attention to the tasks at hand; and the ability to demonstrate to, shape, and instruct each other.

Research shows that teachers promote learning when the social structures are carefully arranged, closely monitored, and adjusted in "real time" to keep students' attention from straying too far from the goals of instruction. We recognize there isn't necessarily a one-size-fits-all formula, but throughout these STEM classrooms there are consistencies and similarities that form the building blocks, or key elements, for teachers to consider.

THE KEY ELEMENTS

In this chapter, we explore in greater detail what the key elements are, how those elements work together to enhance the student learning experience, and the importance of making the learning real-world and relevant to the students. The four key elements are standards, engagement, integration, and assessment.

Standards

The most logical place to start always seems to be at the beginning. When we think about teaching, the first thing that comes to mind is "*What* do we want our students to know and learn?" Sometimes we find ourselves being driven by the demand for higher test scores, so we worry that we should be using checklists to ensure we've covered everything. Or we are swayed by some really trendy topic that's buzzing everywhere in the educational landscape, and we want to jump on the bandwagon. Regardless, at some point we need the guidance of standards if we are going to plan for long-term coherence and deeper student understanding. Standards are the framework that guide us on the *what*. They define what students need to know, describe what is developmentally appropriate, and prescribe what students need to be prepared for in the future.

Standards provide the direction on how to think scientifically, read effectively and efficiently, process and comprehend thoroughly, mathematically solve and explain, and so much more. Standards don't tell us how to teach, or what the best resources to use are, or even which strategies are the most effective. Standards represent the *what* in "What do the students need to know?"

Whether you are in a state that has fully adopted or partially adapted the *Next Generation Science Standards* (*NGSS*; NGSS Lead States 2013), or in a state that has developed its own set of science standards, STEM teaching and learning is still obtainable. Standards are the guide to help you determine what your students need to know and learn. Creating an integrated STEM learning environment will help you tackle the curriculum and organize the individual instructional pieces in a way that ties the concepts together in a meaningful and relevant manner for your students.

Engagement

The standards inform us of the *what*, but that doesn't make the learning appealing or interesting to the students. We now have to get students engaged and eager to participate in the learning of the *what*. When we begin with a scenario that revolves around phenomena, we open the door to students' curiosity and encourage their participation to explore the world around them. According to *Using Phenomena in NGSS-Designed Lessons and Units*,

> *Natural phenomena are observable events that occur in the universe and that we can use our science knowledge to explain or predict. The goal of building knowledge in science is to develop general ideas, based on evidence, that can explain and predict phenomena.* (Achieve Inc. 2016, p. 1)

Part of understanding the world around us is identifying how we are connected to that world. Phenomena are authentic ways to "hook" the learners, to give them something that elicits their curiosity and begs them to ask questions: "Why did that happen?" "How did that happen?" "Why does it work like that?" or "What will happen if . . . ?" Phenomena vary in how long they take students to explain, explore, or model. According to Achieve Inc. (2016), they can be very in-depth and sophisticated, involving critical thought and research, such as "What happened to the aspens when wolves were introduced into Yellowstone Park?" Or they can simply be a puzzling observation of an everyday occurrence or common experience, such as "Why isn't rainwater salty?"

Phenomena can be thought of as both *anchoring* and *investigating* in the course of instruction. As you develop your lessons and units, begin to look for an anchoring phenomenon, something that firmly establishes the overall foundation for student learning that will help inspire curiosity and drive learning across the entire unit. An investigating phenomenon can be the type of observable event or process that students might engage with on a more frequent basis. These phenomena also help to develop the driving question(s) for the investigations. Remember the standards are the *what* that you want your students to learn. Now you must guide their thinking. Consideration for the driving questions that your students will generate, and how you will help them work toward sense making as they investigate phenomena, is an essential part of the *how*.

Don't worry if there are a few detours; it's all part of the process of sense-making for students. Just being told or lectured about a concept does not make it real for the learner. Students need to experience it and connect with it to truly comprehend it. When considering a phenomenon, remember to think about its age appropriateness, especially for K–2 students. Watching a tanker car implode is super cool, but if a student does not yet have the sophisticated scientific knowledge to begin to explain the phenomenon, it won't be much more than a cool trick seen on YouTube. Remember this message seen on a birthday card: "Magic is everywhere if you don't understand science." Throughout this book, we discuss the development and use of phenomena specific to the STEM lessons modeled.

Integration

As K–2 teachers, we know that timing is everything. Don't be the last class in the lunch line, remember to be at the bus loading zone before departure, and find a way to fit it all in within a six-and-a-half-hour day! We're masters at juggling and adapting quickly to the idea of wearing many hats. Our instructional day slowly evolves into a fantastic dance of integration if we hope to address everything we want and need to cover. Because, in between seven or more subject areas that we're responsible for teaching, we've also got specials classes, recess, nurse visits, assemblies, field trips, and many other situations. These other activities, which more often than not truly enhance a child's daily learning experience, do take away from that precious instructional time.

As there are a vast number of definitions of STEM, so too are there multitudes of instructional approaches. But as we offered in our definition (p. 3), and through the lenses we're wearing as K–2 educators, we intentionally look for the key element of integration for our STEM teaching and learning environment. We know as K–2 teachers that it's imperative to get it all in, and research also backs the notion of integration being good for children.

According to a report from the Michigan Department of Education (2014), there are eight benefits to integrated and interdisciplinary teaching:

1. *Coherent conceptual knowledge.* Integrated studies lead to coherence in learning, with the associated development of more elaborated concept development.
2. *Depth of knowledge.* Integration results in the teaching of depth versus breadth, encourages multiple intelligences, and allows for the infusion of literacy and thinking skills among other enhancements.
3. *Opportunities for individualization.* Integrated programs reveal more individualized learning opportunities than traditional methods.
4. *Motivational improvements.* Curriculum integration promotes student involvement and participation in school, and it enhances motivation for academic learning. Integrated models are heralded as an effective alternative to more traditional curriculum-delivered models with the benefit that it heightens curricular relevance.
5. *Enhanced sense of community.* Collaboration is enhanced through the processes of planning and revision of the curriculum. The collaborative meetings bring teachers together and provide a ripe context for professional learning communities and discussions.

6. *Feelings of connection among students.* Integration increases these feelings, as well as the team spirit among the students.
7. *Brain development.* Imaging technologies show that integration results in dendrite growth, allowing for more and better brain connections. Because the brain works in an integrated fashion, integration is in sync with pedagogical best practices.
8. *Overall achievement effects.* Academic performance equals or surpasses the results of students in discipline-based programs. These integrated curricular programs are successful across all four core academic areas (language arts, mathematics, social studies, and science), and at all grade levels.

When considering the above list (which was not specifically written to encompass a STEM teaching and learning environment), it's exciting to see that these are attributes and qualities we hope for and want all students to experience. Something like integration, which K–2 teachers do so often as a normal part of their day—whether it's for survival and to get it all in, or because they've grown enough in their teaching and confidence to experience firsthand how it really does enhance the learning experience—turns out to be a pretty amazing way for children to learn!

Throughout this book, we will show you multiple ways to incorporate different levels of integration, as well as a variety of approaches. It's not a one-size-fits-all thing, but there are some important considerations for planning your integrated STEM units and lessons, regardless of the content area that will be serving as the driver.

Assessment

Assessing students' understanding is a critical component for ensuring they are making connections and developing the skills necessary to successfully tackle more sophisticated concepts as they move along the developmental progressions. Assessment also provides the opportunity to reflect on one's own instruction and lesson planning. It includes both ongoing temperature checks (formative) and overall evaluations (summative), and it must always be explicit if we hope to help students grow and feel empowered with their own learning. It is a very important moment in a teacher's career when she or he realizes the true influence that assessments can have on helping move students forward in their sense-making as opposed to penalizing them for not being able to recall memorized facts for a test.

With STEM teaching and learning, the assessment process also reflects an authentic approach to capturing what students are thinking and understanding. There are concepts and skills that may require a more traditional, paper-and-pencil approach to monitor understanding. Performance-based assessments not only provide insight into what students know, but also allow the students to take ownership of their learning by using that knowledge to solve real-world and relevant issues, questions, challenges, or scenarios.

In an integrated STEM classroom, students are comfortable relying on knowledge they have built from all the content areas and applying that knowledge to construct explanations and design solutions. Students become comfortable integrating their collective knowledge because they have been learning in that manner, which helps them see how all the disciplines connect in the real world and especially in their own lives. These multidimensional assessments provide greater insight into the strengths and weaknesses of the students' thinking and can help identify areas for further practice.

There are many ways to infuse performance-based assessments into your STEM classroom. The performance-based task is where the students' focus is on their ability to demonstrate their knowledge and skills. Remember that you're providing your students with the opportunity to apply what they have been learning. Memorizing facts for a test is not application. Obviously not all students just memorize—they are learning and can competently pass a test—but without an opportunity to actually apply what they are learning, there is no guarantee they have made a connection to the content and have made the personal connections necessary for sense-making. We all remember a science project we worked on, a soap-box derby car we designed, an apple pie we baked for the fair, or a vest we sewed in home economics class. It was the opportunity to apply what we were learning that had staying power and made the learning and lessons relevant and memorable. Even if the final outcome wasn't perfect, that chance to show what we were thinking made a lasting impression.

In the units and lessons modeled throughout this book, we share a few examples of performance-based assessments that include problem-based (similar to project-based) learning and engineering design challenges. You'll see in these lessons, regardless of the approach, that students are provided with a framework for completing the assessment that includes the setting and purpose.

THE KEY ELEMENTS IN ACTION

Once we have the key elements addressed and in place as we are planning, we can begin to consider what it all looks like in action. How will we engage students with the phenomena and sufficiently excite them to generate their own questions? How will that enthusiasm help them tackle and address our guiding questions? What, then, are the next steps to get this plan into action?

In 2012, the National Research Council (NRC) published *A Framework for K–12 Science Education*. Shortly after, in 2013, the *Next Generation Science Standards* (NGSS Lead States 2013), which are based on the *Framework*, outlined a broad set of expectations for K–12 students in science and engineering. These documents look at how students learn, as well as how to improve science education within three dimensions of learning: science and engineering practices, crosscutting concepts, and disciplinary core ideas (physical science, life science, and Earth and space science).

For the first time, these two documents address what science education should look like and how each dimension must be integrated into a set of standards, curricula, instructions, and assessments if we are to truly support students' meaningful learning in science and engineering. As we focus more on integration among the STEM disciplines, we begin to find the NRC research applicable and helpful when we also tie in the practices found in language arts and mathematics.

The natural connections in the concepts and skills are made throughout the Common Core State Standards for Mathematics and English Language Arts as well as within the *NGSS*. Recognizing those natural connections confirms the idea that integrated teaching and learning is beneficial for learners as they work to make sense of the world around them (NRC 2012).

As we look to our standards for all content areas, we are identifying what we want our students to know. When we start to consider what that means for our K–2 students, we have to imagine all the nuances that make them such wonderful little people and how we will get them to listen, pay attention, focus, behave, learn, follow along, try things out, share, cooperate, and so on.

Let's go back to where we started in this chapter. Make the learning all about them. When they are the center of it all, children rarely ignore what's going on. Rarely can a student just sit and get the information. Students have to be able to make a personal connection and to identify what it has to do with them. Sometimes it is easy for a child to see, other times it's more abstract, and without the opportunity to interact with it, understanding becomes challenging and complicated. And dare we say the "F" word? *Fun!* School should be fun for students, of all ages, and especially for our K–2 learners. There is a lot out there to compete with these days, but by inviting your students in to get dirty (OK, maybe not literally) and have fun, your classroom becomes a place where they want to be. The science and engineering practices will help you engage them with the content like scientists, technologists, engineers, and mathematicians.

The science and engineering practices are a true reflection of the behaviors that real scientists and engineers engage in as part of their work. While we don't expect the same level of sophistication from our students as those professionals in their respective fields, we do want to engage our students authentically so they can begin to ask questions, define problems, construct explanations, and design solutions about the exciting world in which they live. Ultimately our goal is to help all students become literate and comfortable, navigating all the STEM disciplines.

Table 3.1 (p. 30) shows a box from the *Framework* (NRC 2012) that distinguishes the practices in science from the practices in engineering. The practices are not linear but instead should be used iteratively and often in combination with one another. Don't create a "completed" checklist or panic if your students don't engage with every practice during every lesson. The goal is to remember that the practices are the "tools" your students will use. They are what the students are doing as they engage with the phenomena and explore the scientific and engineering core ideas, resources, manipulatives, and literature. The practices are just as the name implies—an opportunity for students to *practice* just like scientists and engineers do in the real world.

Table 3.1. The science and engineering practices

1. Asking Questions and Defining Problems	
Science begins with a question about a phenomenon, such as "Why is the sky blue?" or "What causes cancer?" A basic practice of the scientist is formulating empirically answerable questions about phenomena to establish what is already known and to determine what questions have yet to be satisfactorily answered.	**Engineering** begins with a problem that needs to be solved such as "How can we reduce the nation's dependence on fossil fuels?" or "What can be done to reduce a particular disease?" or "How can we improve the fuel efficiency of automobiles?" A basic practice of engineers is to ask questions to clarify the problem, determine criteria for a successful solution, and identify constraints.
2. Developing and Using Models	
Science often involves the construction and use of models and simulations to help develop explanations about natural phenomena. Models make it possible to go beyond observables and simulate a world not yet seen. Models enable predictions of the form "if ... then ... therefore" to be made in order to test hypothetical explanations.	**Engineering** makes use of models and simulations to analyze extant systems and to identify flaws that might occur, or to test possible solutions to a new problem. Engineers design and use models of various sorts to test proposed systems and to recognize the strengths and limitations of their designs.
3. Planning and Carrying Out Investigations	
Scientific investigations may be conducted in the field or in the laboratory. A major practice of scientists is planning and carrying out systematic investigations that require clarifying what counts as data and conducting experiments that identify dependent and independent variables.	**Engineering investigations** are conducted to gain data essential for specifying criteria or parameters and to test proposed designs. Like scientists, engineers must identify relevant variables, decide how they will be measured, and collect data for analysis. Their investigations help them to identify the effectiveness, efficiency, and durability of designs under different conditions.
4. Analyzing and Interpreting Data	
Scientific investigations produce data that must be analyzed in order to derive meaning. Because data usually do not speak for themselves, scientists use a range of tools—including tabulation, graphical interpretation, visualization, and statistical analysis—to identify the significant features and patterns in the data. Sources of error are identified and the degree of certainty calculated. Modern technology makes the collection of large data sets much easier, providing secondary sources for analysis.	**Engineering investigations** include analysis of data collected in the tests of designs. This allows the comparison of different solutions and determines how well each one meets specific design criteria—that is, which design best solves the problem within the given constraints. Like scientists, engineers require a range of tools to identify major patterns and interpret the results. Advances in science make analysis of proposed solutions more efficient and effective.

Continued

Table 3.1 (*continued*)

5. Using Mathematics and Computational Thinking	
In science, mathematics and computation are fundamental tools for representing physical variables and their relationships. They are used for a range of tasks such as constructing simulations; statistically analyzing data; and recognizing, expressing, and applying quantitative relationships. Mathematical and computational approaches enable prediction of the behavior of physical systems along with testing of such predictions. Moreover, statistical techniques are also invaluable for identifying significant patterns and establishing correlation relationships.	**In engineering,** mathematical and computational representations of established relationships and principles are an integral part of the design process. For example, structural engineers create mathematically based analyses of designs to calculate whether they can stand up to expected stresses of use and if they can be completed within acceptable budgets. Moreover, simulations provide an effective test bed for the development of designs as proposed solutions to problems and their improvement, if required.
6. Constructing Explanations and Designing Solutions	
The goal of science is the construction of theories that provide explanatory accounts of the material world. A theory becomes accepted when it has multiple independent lines of empirical evidence, greater explanatory power, a breadth of phenomena it accounts for, and explanatory coherence and parsimony.	**The goal of engineering** is a systematic solution to problems that is based on scientific knowledge and models of the material world. Each proposed solution results from a process of balancing competing criteria of desired functions, technical feasibility, costs, safety, aesthetics, and compliance with legal requirements. Usually there is no one best solution, but rather a range of solutions. The optimal choice depends on how well the proposed solution meets criteria and constraints.
7. Engaging in Argument From Evidence	
In science, reasoning and argument are essential for clarifying strengths and weaknesses of a line of evidence and for identifying the best explanation for a natural phenomenon. Scientists must defend their explanations, formulate evidence based on a solid foundation of data, examine their understanding in light of the evidence and comments by others, and collaborate with peers in searching for the best explanation for the phenomenon being investigated.	**In engineering,** reasoning and argument are essential for finding the best solutions to a problem. Engineers collaborate with their peers throughout the design process, with a critical stage being the selection of the most promising solution among a field of competing ideas. Engineers use systematic methods to compare alternatives, formulate evidence based on test data, make arguments to defend their conclusions, critically evaluate the ideas of others, and revise their designs in order to identify the best solution.
8. Obtaining, Evaluating, and Communicating Information	
Science cannot advance if scientists are unable to communicate their findings clearly and persuasively or learn about the findings of others. A major practice of scientists is thus to communicate ideas and the results of inquiry—orally; in writing; with the use of tables, diagrams, graphs, and equations; and by engaging in extended discussions with peers. Science requires the ability to derive meaning for scientific texts (papers, the internet, symposia, lectures, etc.), to evaluate the scientific validity of the information thus acquired, and to integrate that information into proposed explanations.	**Engineers** cannot produce new or improved technologies if the advantages of their designs are not communicated clearly and persuasively. Engineers need to be able to express their ideas—orally; in writing; with the use of tables, graphs, drawings, or models; and by engaging in extended discussions with peers. Moreover, as with scientists, they need to be able to derive meaning from colleagues' texts, evaluate the information, and apply it usefully.

Source: Adapted from NRC (2012).

A Framework for K–12 Science Education (NRC 2012) calls for a progression of the practices in Table 3.1 across the K–12 continuum. It provides a broad description of what students should be able to do for each practice at the end of the particular grade band: K–2, 3–5, 6–8, and 9–12. According to the *Framework*,

> *The progression for practices across the grades follows a similar pattern, with grades K–2 stressing observations and explanations related to direct experiences, grades 3–5 introducing simple models that help explain observable phenomena, and a transition to more abstract and more detailed models and explanations across grades 6–8 and 9–12.* (NRC 2012, p. 34)

When sharing with teachers who are new to the practices, we often suggest that they think of it as taking the words we so often use in the elementary grades—*hands-on* and *minds-on*—and being very deliberate and specific about what the students are actually doing in science and engineering. Imagine how empowered you will begin to feel when you can specifically identify and explain what your students are using their hands and minds for! The lesson models in this book will suggest where the practices come into play for students and give examples of what they are doing while engaging with the STEM disciplines.

REFERENCES

Achieve Inc. 2016. Using phenomena in NGSS-designed lessons and units. September. *www.nextgenscience.org/sites/default/files/Using%20Phenomena%20in%20NGSS.pdf.*

Michigan Department of Education. 2014. *Curriculum integration research: Re-examining outcomes and possibilities for the 21st century classroom.* Lansing, MI: Department of Education, Office of Education Improvement and Innovation.

National Research Council (NRC). 2012. *A framework for K–12 science education: Practices, crosscutting concepts, and core ideas.* Washington, DC: National Academies Press.

National Research Council (NRC). 2014. *STEM integration in K–12 education: Status, prospects, and an agenda for research.* Washington, DC: National Academies Press.

NGSS Lead States. 2013. *Next Generation Science Standards: For states, by states.* Washington, DC: National Academies Press. *www.nextgenscience.org/next-generation-science-standards.*

CHAPTER 4

TACKLING THE CORE INSTRUCTIONAL TIME

The more that you read, the more things you will know.
The more that you learn, the more places you'll go.

—Dr. Seuss, *I Can Read with My Eyes Shut!* (1978)

In this chapter, we consider how the STEM (science, technology, engineering, and mathematics) classroom can be used in concert with a core reading program to achieve instructional goals and objectives for all students.

THAT SACRED TIME

It doesn't matter in which part of the United States you teach, whether you're in an urban, suburban, or rural environment, in either a public or a private setting—if you teach K–2 students, you spend a significant part of your day immersed in language arts instructional practices. Often in the morning when our young learners are most awake and hopefully focused, the majority of instructional time centers on developing their literacy skills. Whether it's called the reading block, language arts time, small- or whole-group reading, literacy instruction, or any other name, it's consistent in its overall intent, structure, goals, and outcomes.

This is the time that K–2 students spend, almost exclusively, learning to read—mastering all the intricacies such as phonics, intonation, fluency, comprehension, spelling, sentence structure, and so much more that it's almost enough to make your head spin. And in some cases, when children are struggling to crack that elusive code and make successful progress, they too may feel like their heads are spinning, trying to make sense of what appears to be senseless. They begin feeling like an outsider to a very exclusive club. In a short amount of time, that feeling begins to cross over into other areas of the school day too. Without the important foundational skills that we call reading, a student begins to turn away from other content-area "clubs" as well.

Transfer of Knowledge

As an experienced elementary educator, and from her many years of working with students and teachers, coauthor Jen Gutierrez has come to understand the transfer of knowledge as one of the principal goals of learning in school. Students should be able to take the knowledge in one context and apply it in another. "Explained = retained," as a teacher shared on a webinar that Jen was delivering. However, students must engage with the concepts in meaningful ways for that transfer to occur. Instruction needs to be relevant, real-world, and rigorous. Students need numerous opportunities to practice, apply, discuss, discover, and connect. Elementary teachers struggle constantly with the challenge of how to fit it all in on a daily basis. Sometimes it feels like the instructional window is quickly closing with the multiple, and at times opposing, curricular demands. And that doesn't even include the unexpected moments that occur, some of which are instructionally supportive, such as an assembly or a special guest.

Educators setting up an instructional model that takes into consideration all the nuances of an elementary school classroom and the children that inhabit it—with the idea of engaging those students in relevant, real-world, and rigorous learning opportunities—would welcome the notion that integrated teaching and learning could effectively help them fit it all in. Using the STEM disciplines to drive the instruction would support the struggling classroom teacher in covering the many standards while at the same time giving students the opportunity to read for the purpose of learning. It is those critical foundation skills that are built as students are learning to read that help them begin to make sense of the words they master as they read to learn. The different content areas can provide young learners with the purpose for learning to read, while unlocking the doors to a world of unlimited possibilities.

In January 2012, Stanford University hosted the Understanding Language Conference, which presented a number of commissioned research papers dealing with understanding the impact that language has on learning. The Understanding Language Initiative is an organization that works with teachers, researchers, and policymakers to further a national dialogue to address literacy and language issues. In one paper, Quinn, Lee, and Valdes (2012) focused on the language demands necessary for success with the *Next Generation Science Standards* (*NGSS*; NGSS Lead States 2013). There were three notable areas in the paper concerning the shifts in science instruction:

1. *The science and engineering practices (SEPs) involve both scientific sense-making and language use.* Four of these SEPs—#2, Developing and Using Models; #6, Constructing Explanations and Developing Solutions; #7, Engaging in Argument From Evidence; and #8, Obtaining, Evaluating, and Communicating Information—focus on student sense-making and reflect a shift in classroom instructional practices that is not familiar to most teachers. "Particularly in the lower grades the activity often ends at the stage of recording observations, with minimal attention paid to interpreting them and almost no attention to constructing models or explanations and refining them through argumentation for evidence" (Quinn, Lee, and Valdes 2012, p. 33). These four SEPs are interconnected and require classroom discourse. As students read, write, draw, and develop their models, they need to speak and listen to the ideas presented by others. They need to engage in forms of argument, obtain and evaluate new information, develop the skills for active listening and using description with precision, and build confidence in their ability to articulate their ideas without a focus on grammatical correction or pronunciation

errors. The SEPs provide that bridge to building strong literacy skills in young learners if teachers focus on explicit instructional strategies that zero in on them.

2. *Science experiences represent an intersection of "doing" and "talking."* The types of collaborative activities described in the new standards require more of students in their use of literacy skills. Young learners engage in experiences that require them to share their thoughts, expressing abstract ideas with language skills that are just beginning to emerge. At the same time, it demands that they make use of a growing facility of common and technical vocabulary to transfer that learning to different applications and situations. These new features of science instruction provide different conventions of discourse. Teachers need to support all "students in developing an understanding of the forms of this discourse as well as those used in written science text" (Quinn, Lee, and Valdes 2012, p. 37).

3. *If students are to succeed in the goals of the new standards, teachers need to adapt instructional strategies that support the acquisition of both science learning and language literacy skills.* This is the case whether teachers are working with the *Common Core State Standards* or the *NGSS*. "[E]ffective teachers incorporate reading and writing strategies in their instruction to promote both science learning and literacy development for all students" (Quinn, Lee, and Valdes 2012, p. 38). There are multiple strategies that can be intentionally employed to help young learners make connections across their learning landscape. Some of those strategies include explicit instruction of reading strategies for scientific texts, prompting students to use the academic language found in the standards, requiring writing in science genres and exemplifying good examples of such writing, encouraging reading of science trade books, and using journal prompts that focus on investigative protocols such as "I observed …," "I predicted …," and "I analyzed …."

All students are expected to engage with scientific phenomena and learn how to describe, predict, explain, and apply. It is important to help them learn how to "decode" the complex sentences of scientific texts, to recognize the context of the vocabulary, and to develop precision in their writing and explanations. (See the "Cracking the Code" box, p. 36.)

Cracking the Code

I have two sons, both grown and flourishing well in the world. They went with me to the school I taught at, and I loved every minute of having them nearby.

What I found very quickly is that they could not be more different, in almost every way, especially when it came to their learning-to-read journey. My youngest, Cody, walked into kindergarten and went right to work, reading and writing without any struggles. He was exceptionally shy as a baby and toddler, so I was delighted to see him blossom.

My oldest, Kyle, came out of my womb talking and would carry around a phone book or dictionary and tell us he was reading (I think because they were the biggest books he could find in the house). He loved everything there was to love about books, and we read together every night. It was our special time and one that could never be missed. And don't even think about skipping pages or making things up. He knew each of those hundreds of beloved books by heart.

After his first day of kindergarten, Kyle was not pleased. I had seen him throughout the morning and he seemed happy. He soon explained to us that no one had taught him how to read that day. Well, as it turned out, he struggled desperately to crack the code.

We worked with Kyle, I got him extra help, I even had him tested. I was at a loss. I could help other people's children learn to read, and do it well, but I could not help my own child.

Finally, in fourth grade, it happened! With the help of an incredible teacher who continued to support his love of learning and all things books, and who did not ever let him think he couldn't do it, Kyle cracked the code and began reading. By the end of the year he was back on track and succeeding well and even on grade level! So much of what was going on in his fourth-grade class centered on real-world connections and why they were relevant to the students. He never lost interest or felt left out because he was still invited to be a part of the learning.

Like every teacher, I encountered myriad new ideas, fads, and buzzwords around education throughout the years. Sometimes I experienced a shift in my instructional practices and sometimes it was just a matter of riding out the wave.

Regardless of the newest curriculum or recent instructional strategy, I realized that none of my students learned in the same way. But what they almost all consistently had in common was the love of a good book. Reading to my students held as much joy as it did with my boys at home, and it was the one way I knew I could help them connect to the learning in a way that was meaningful, fun, and personal.

Providing a learning environment that invites children in while supporting their own growing skills is key to building a teaching and learning space that gets students to engage in that transfer that is essential to making sense.

—Jen Gutierrez

The Language Arts Instructional Block

When we are working with our students in small-group instruction or support, we are primarily focused on word recognition and its components, phonological awareness, decoding (and spelling), and sight recognition. We work with our students to help them develop a level of automaticity as they become more sophisticated with their skills. If there are red flags and students are struggling, then we must also provide additional support to help them master these components. This chapter is not meant to diminish the intense time, preparation, skill, and instruction it takes to work with students in this instructional setting. As educators we also acknowledge that there are a multitude of successful instructional practices and strategies being implemented to strengthen students' progress in learning to read. Learning to read is key. However, having a reason to read is what is inspiring about incorporating STEM learning into the classroom. STEM experiences are about the application of what the students are learning. The focus of this chapter is to provide support for infusing this STEM learning-to-read journey while reading to learn in an authentic, relevant setting.

While we are helping students build word recognition, we are also helping them become increasingly strategic with their language comprehension, background knowledge, vocabulary knowledge, language structures, verbal reasoning, and literacy knowledge. Our ultimate goal is helping students become skilled readers with fluent execution and with coordination of word recognition and text comprehension (Scarborough 2001). As students begin to make connections with language in a meaningful context, not only are they engaged in new and maybe unfamiliar content, but they also are working on sense-making as they begin to think, read, write, speak, and listen like a scientist, technologist, engineer, or mathematician.

The language arts instructional block is a wonderful opportunity to help students on their learning-to-read path because it doesn't separate them from the reading to learn that is simultaneously occurring. At the end of first grade, oral vocabulary is a significant predictor of comprehension 10 years later (Cunningham and Stanovich 1997). Our students need opportunities for engaging in meaningful conversations with their peers about the world around them. Giving students something important to talk about is the bedrock of engaged, excited, inquisitive, and curious children. Giving a daily journal prompt that seems canned or repetitive will lose its power to motivate very quickly.

Designing learning opportunities that are centered on academic language and the formal communication structure and words that are common in books and in school, and built around the STEM disciplines, means that our students are engaging in many opportunities that support the growth of academic language skills. Those skills include inferential language skills, narrative language skills, and academic vocabulary knowledge. Using the STEM disciplines as a driver means we're engaging students in conversations before, during, and after read-alouds, which encourages higher-level thinking. We're using open-ended questions to challenge students to think about the message and how they can apply it to the world around them. And we're modeling for our students how to provide answers that fully address the questions and illustrate levels of critical thinking.

As we help students develop their narrative language skills and the ability to organize information in logical sequences, as well as connect that information using appropriate complex grammatical structures, we can also support them by scaffolding their responses. As students become more comfortable with the given element, they will require fewer prompts and modeling and will begin using narrative structure elements independently.

Vocabulary is always a hot topic in the elementary classroom. Word lists for memorizing terms, the recalling of definitions, and spelling tests to check student learning all exist outside a context that could help students connect with their knowledge and understanding in a way that makes sense. These tasks, while structured and repetitive, are usually not as effective as we intended for building long-term understanding. Academic vocabulary knowledge is the ability to comprehend and use words and grammatical structures common to formal writing. Students don't just need academic vocabulary knowledge, though; they need skills to use this vocabulary in authentic contexts. As our lessons shift to include more real-world, authentic connections for students, we are building in a perfect opportunity for the students to engage in conversations that will help them develop their oral fluency.

According to *STEM Integration in K–12 Education*, "for integrated STEM education to be successful, students need to be able to move back and forth between the acquisition of disciplinary knowledge and skill in their applications to problems that call on competencies from multiple disciplines" (NRC 2014, p. 71). Students who participate in collaborative conversations in conjunction with lab experiences or engineering challenges will likely show more detail or depth of knowledge in their written work.

Any one of the STEM disciplines used as the driver for a lesson will provide those text connections where students can also practice their word recognition and foundational skills through reading, writing, listening, and speaking activities that are not only content driven, but interesting and engaging for the learners.

REFERENCES

Cunningham, A. E., and K. E. Stanovich. 1997. Early reading acquisition and its relation to experience and ability 10 years later. *Developmental Psychology* 33 (6): 934–945.

National Research Council (NRC). 2014. *STEM integration in K–12 education: Status, prospects, and an agenda for research.* Washington, DC: National Academies Press.

NGSS Lead States. 2013. *Next Generation Science Standards: For states, by states.* Washington, DC: National Academies Press.

Quinn, H., O. Lee, and G. Valdes. 2012. Language demands and opportunities in relation to *Next Generation Science Standards* for English language learners: What teachers need to know. Paper presented at the Understanding Language Conference, Stanford University, January.

Scarborough, H. 2001. Connecting early language and literacy to later reading (dis)abilities: Evidence, theory, and practice. In *Handbook of early literacy*, ed. S. B. Neuman and D. K. Dickinson, 97–110. New York: Guilford Press.

Using the W.H.E.R.E. Model Template

"Would you tell me, please, which way I ought to go from here?" asked Alice.
"That depends a good deal on where you want to get to," said the Cat.
"I don't much care where—" said Alice.
"Then it doesn't matter which way you go," said the Cat.

—Lewis Carroll, *Alice in Wonderland* (1865)

This classic quote by Lewis Carroll is very appropriate for our conversation about developing your own STEM (science, technology, engineering, and mathematics) units. We as educators know that understanding is revealed when students make sense of and transfer their learning through authentic performance tasks. This is supported by the research of Grant Wiggins and Jay McTighe (2005) in *Understanding by Design*. And what could be more authentic than developing problem-based STEM lesson experiences?

In this chapter, we introduce the research-based W.H.E.R.E. model, which is described in the book *STEM Lesson Guideposts* (Vasquez, Comer, and Villegas 2017). The W.H.E.R.E. model template presents a clear and actionable process for curriculum developers and classroom teachers to follow as they set out to develop their own 21st-century STEM experiences. The W.H.E.R.E. model will help educators and curriculum leaders weave their own hands-on, inquiry-focused experiences with more relevant and rigorous tasks using their own content while implementing new standards, practices, and crosscutting concepts as they create STEM units.

As with any curricular development work, it is important to keep in mind that the W.H.E.R.E. model template is designed to assist educators with the full planning of their STEM experience, providing thoughtful reminders of key instructional considerations. The W.H.E.R.E. model template is described below:

- **W** = *What* needs to be learned? And *Why*?
- **H** = *How* do I plan to get there? (What is the general roadmap?)
- **E** = What *Evidence* of learning will be used, and how will I *Evaluate* the final product or project?

- **R** = How will I provide opportunities to increase the *Rigor* and *Relevance*?
- **E** = How do I *Excite* and *Engage* my students and allow for *Exploration*?

An effective STEM unit embodies these facets as essential goals of its instruction. The template from *STEM Lesson Guideposts* (Vasquez, Comer, and Villegas 2017) captures these ideas within the individual guideposts that support effective planning. To this end, we have adapted and expanded the Understanding by Design (UbD) model, a three-stage backward design process developed by Wiggins and McTighe (2005). Their planning highlights three key stages in unit development:

1. **Identify desired results**
    - What should students be able to know, understand, and do?
    - What content and skills will be necessary for student understanding?
    - What are the big ideas, key concepts, knowledge, and skills that describe what the students will know and be able to do?
2. **Determine acceptable evidence for assessment**
    - How will you know whether your students have achieved the desired results?
    - What will serve as evidence of student understanding and proficiency?
3. **Plan the learning experiences**
    - What prior knowledge and skills will the students need to perform successfully?
    - What activities foster integration?
    - What levels of integration will be most effective to accomplish the learning goals?
    - How will there be opportunities for all students to participate?
    - What resources and materials will be needed to accomplish the goals?

We have expanded these three stages into the five guideposts found in the W.H.E.R.E. model.

THE W.H.E.R.E. MODEL TEMPLATE

The W.H.E.R.E. model template (see Figure 5.1) provides a working framework for planning STEM units by identifying strategic guiding questions and helping you navigate the five key areas—your "guideposts"—for planning an effective STEM unit. Each guidepost is introduced with a question that frames the purpose of that section and provides direction for the information that should be developed there.

W = What and Why: What needs to be learned and why?

The **W** guidepost represents the *what* and the *why* in planning the instruction (see the first part of Figure 5.1). This section reflects Stage 1 in the UbD model above, focusing on the desired results of the learning experience—the *what* and, relatedly, the *why*. This guidepost helps focus the

Figure 5.1. The W.H.E.R.E. model planning template

<table>
<tr><td>W</td><td colspan="3">What are the desired results, including the big ideas, content standards, knowledge, and skills?
• List the content standards and what the students will know and be able to do.</td><td colspan="3">Why would the students care about this knowledge and these skills?
• Craft the driving question that will lead to the development of the integrated tasks that provide for the application of the content, knowledge, and skills.
• List the essential questions that can be answered as a result of the learning.</td></tr>
<tr><td>H</td><td colspan="6">How do I plan to meet this goal?
• Identify the pathway, including the major tasks and milestones that result in answering the driving question.</td></tr>
<tr><td>E</td><td colspan="2">What Evidence of learning will be used, and how will I Evaluate the final product or project?
Preassessment. What prior knowledge is needed for this task?
• Identify the prerequisite skills and understandings.</td><td colspan="2">Formative. How will I measure student progress toward understanding?
• Establish the assessment tools you will use to monitor progress and inform instruction.</td><td colspan="2">Summative. What criteria are needed for students to demonstrate understanding of the standards, content, and skills?
• Create a checklist of criteria for use in a rubric.</td></tr>
<tr><td>R</td><td colspan="3">Rigor
How can I increase students' cognitive thinking?
• Identify tasks that can elevate student thinking, improve inquiry, and increase conceptual understanding.</td><td colspan="3">Relevance
Does the learning experience provide for relevant and real-world experiences?
• Identify current topics and local issues that can make the tasks more engaging.</td></tr>
<tr><td>E</td><td colspan="2">Excite
What is the hook to excite the learner?
• Create the scenario to engage the learner.</td><td colspan="2">Engage
How will students be cognitively engaged throughout the unit?
• List the STEM practices that will be used as evidence.</td><td colspan="2">Explore
What activities will help students address the driving question?
• List questions for students to investigate that will lead them to a deeper understanding of the content and skills.</td></tr>
</table>

Source: Vasquez, Comer, and Villegas (2017, p. 13).

instructional planning on the critical elements of applying content knowledge and skills to the real world. It provides an opportunity to tease apart the performance expectation of the *Next Generation Science Standards* and highlight each of its composite dimensions. What is the disciplinary core idea being developed? How will students interact with this core idea and how will they connect it to the broader concept? This stage also hones the driving question, which should capture and communicate the purpose of the learning.

Strategic guiding questions for *what* and *why* are as follows:

- What are the desired results, including the big ideas, content standards, knowledge, and skills (e.g., 21st-century skills), that this lesson or unit will teach?
- Why would the students need to know and understand these concepts and skills?
- What are the essential questions that will lead to student understanding of the content?
- What is the driving question that will lead to the development of integrated tasks that provide the application of the desired content, knowledge, and skills?

H = How: How do I plan to meet this goal?

The **H** guidepost represents the general development pathway that the unit will follow (see the second part of Figure 5.1, p. 41). How do you plan to meet this goal? What are the basic core experiences that the students will engage in to demonstrate their understanding of the disciplinary core idea? This guidepost helps you identify your planning needs, calling out the high-level learning milestones that students should meet as they progress through the unit. It is a place to organize the sequence of learning experiences that will scaffold their understanding to meet those milestones. You will develop the specifics for each of these later in the template, by adding in more refined details in the Exploration section of the template. The *H* guidepost also provides an opportunity to begin thinking about and considering what prerequisite knowledge or skills students would need for each of these learning milestones.

Strategic guiding questions for *how* are as follows:

- What learning experiences will enable students to understand the concepts and skills in meaningful ways?
- How will the sequence of these learning experiences help students to construct their understanding and apply the skills?
- What are the major tasks or milestones that will lead toward answering the driving question?

E = Evidence and Evaluate: What evidence of learning will be used, and how will I evaluate the final product?

The first **E** guidepost represents the development of the various assessments that will be used to monitor students' progress and their overall success (see the third part of Figure 5.1, p. 41). In following the UbD model, knowing what you will assess and how you think about the delivery of that assessment will help you plan the instruction to meet those desired goals. Think about the final product the students will produce that will demonstrate their learning and understanding. What form will it take? Is it a project, a report, an oral presentation, or an evidence-based scenario task? Will it be an individual endeavor or a collaborative, small-group project? What tool will you need to evaluate their learning?

These thoughts will provide guidance as you develop your lessons. To achieve those outcomes, what intermediary knowledge and understandings will students need to have and be able to demonstrate? These reflections can help you formulate your formative assessment sequence. District grading policies also come into play to help determine the method and frequency of your assessments. This section of the template emphasizes the importance of the assessment tools prior to actual lesson planning.

Strategic guiding questions for *evidence* and *evaluate* are as follows:

- What prior knowledge and skills will the students need to have and be able to demonstrate?
- What formative assessments will be used to measure student progress toward understanding and to inform instruction?
- What summative assessment, culminating product, or task will demonstrate the students' understanding of the standards, content, and skills?
- What criteria and tools will be used for assessing student success?

R = Rigor and Relevance: How will I provide opportunities that will lead to increased rigor and relevance?

The **R** guidepost reflects the research of Willard Daggett (2008), drawing on the Rigor and Relevance Framework developed by the International Center for Leadership in Education (see the fourth part of Figure 5.1, p. 41). Focusing on rigor and relevance shifts the instructional focus from "teacher-led" to "student-driven," and it moves the perspective from a classroom context to a real-world setting. The *R* guidepost emphasizes the different factors that can increase a student's cognitive involvement while at the same time making the learning experience more meaningful.

Strategic guiding questions for *rigor* and *relevance* are as follows:

- How will the learning experiences extend the students' thinking?
- What intriguing questions will foster greater inquiry to increase conceptual understanding?
- What opportunities will provide for increased relevance and real-world connections for the students?

E = Excite, Engage, Explore: What exploration activities will excite the students and engage them in the STEM practices?

The second **E** guidepost focuses on ways to engage and excite students as they navigate the lessons that make up the unit (see the fifth part of Figure 5.1, p. 41). In the UbD model, Stage 3 identifies the importance of planning these learning experiences with an emphasis on engaging student interest and addressing the needs of all the learners in the classroom. In the W.H.E.R.E. model, this section also focuses on moving beyond the physical and social engagement of the classroom experience and highlights the importance of cognitive engagement.

Cognitive engagement occurs when students are asking questions, applying knowledge in new ways, and self-assessing their own learning. The STEM practices reflect this kind of cognitive engagement. The exploration tasks lead students to the desired learning outcomes and tie together the other components of the W.H.E.R.E. model. These activities should address the essential questions in the Why section, which in turn lead to deeper student understanding.

Strategic guiding questions for *excite*, *engage*, and *explore* are as follows:

- What is the scenario that will "hook" the learner and foster engagement?
- How will the students be cognitively engaged?
- Which STEM practices will the students engage in?
- How will the explorations provide evidence of this engagement?
- How do the explorations address the driving question?

USING THE PLANNING TEMPLATE

The W.H.E.R.E. planning template (Figure 5.1, p. 41) is not designed as an exercise to complete from top to bottom, filling in each section before moving to the next. Rather, consider the template as a "placemat" on which you can record and organize your thoughts during the process of brainstorming as you develop your own STEM unit. The template is both an organizational structure and a reflective observation tool. As an organizational structure, the W.H.E.R.E. model allows you to think, rethink, revise, and reorganize your thoughts as you plan the unit.

Thinking of the *why* may give you ideas on how to engage the learner, which is found in the second *E* guidepost. Thinking about a formative assessment task in the first *E* guidepost can spark ideas on how to improve an exploratory investigation in the second *E* guidepost, which might lead to ideas on how to make it more relevant in the *R* guidepost. In other words, the template is a mosaic of planning, filled in as ideas for each section arise in your thinking; yet once completed, it provides a thorough picture of the unit as a whole. It's important to remember that the template needs to be considered as a whole, rather than just focusing on one section at a time.

Also, the work in the W.H.E.R.E. template should never be considered finished. Once the unit is delivered, the completed template should serve as a reflective piece in which to look back at the instruction in the unit. Did the students gain the understanding that was intended? Were the investigations and student experiences rich enough and challenging enough to keep them interested? Is there anything new, local, or more current that you can add? Did your assessments provide the kind of insight you needed to guide future learning or counter misconceptions? Use the template to review what worked well, what needed improvement, or what alternative materials might be incorporated to make it even better. The unit should be reviewed each year so that it reflects local or current phenomena that can serve to energize and motivate your students. The completed W.H.E.R.E. model template becomes a living document that you and your colleagues can share for many years.

CHOOSING THE DRIVER

A word of caution in preparing your STEM unit: It is important to be aware of a few missteps in developing the integrated learning experiences. An overstuffed STEM unit can create confusion and disinterest in students. Cognitive research describes the point of diminishing return when too much is asked too soon. Remember that the learner is processing multiple sources of information at any given time, trying to separate, sort, store, and manage the relevant pieces of information and experiences necessary to make sense. Dividing learners' attention between competing sources of input can create an obstacle to their ability to process anything.

As it says in *STEM Integration in K–12 Education*,

> *These aspects of cognition point to a potential drawback of integration: without effective guidance, the effort to make connections among multiple disciplines in the context of a complex problem or situation could overwhelm students and inhibit learning. Design of the integrated experiences must balance the richness of the integration and real-world contexts against the constraints of cognitive demands of processing information that is separated in time, space, or across disciplines and types of representations.* (NRC 2014, p. 84)

In other words, careful planning should focus on one or two main ideas that student should know and understand at the completion of the learning cycle.

In planning the STEM unit, we use the analogy of a four-passenger car. There is only one driver in the car. The driver takes control of how the vehicle operates and where the people go. There can be a front-seat passenger. This copilot provides additional direction and helps in navigating everyone to the destination. In the back seat are the other passengers. They are just along for the ride, now and then adding commentary to the trip. They don't steer the car or determine the destination. In creating your STEM unit, we recommend one discipline to be the "driver." It is the main focus of the learning. A second discipline, the copilot, can directly support that learning. This front-seat passenger helps navigate the learning and ensures that we get to the destination on time. Any other disciplines we add should only be tangential to the learning. They are the back-seat passengers. They come along for the ride, and add color commentary and context to the STEM experience, but are not influential in getting to the destination.

The old adage, "Too much of a good thing is too much," applies to learning also. Decide what the essential driver of the learning experiences will be and plan to focus on that. The addition of secondary or supporting "passengers" that could enhance the overall experience and not complicate it is an option that can be added when necessary.

REFERENCES

Daggett, W. R. 2008. *Rigor and relevance from concept to reality.* New York: International Center for Leadership in Education.

National Research Council (NRC). 2014. *STEM integration in K–12 education: Status, prospects, and an agenda for research.* Washington, DC: National Academies Press.

Vasquez, J. A., M. Comer, and J. Villegas. 2017. *STEM lesson guideposts: Creating STEM lessons for your curriculum.* Portsmouth, NH: Heinemann.

Wiggins, G., and J. McTighe. 2005. *Understanding by design.* 2nd ed. Alexandria, VA: Association for Supervision and Curriculum Development.

Developing a STEM Unit With Math as the Driver—Straw Bridges

A square was sitting quietly
Outside his rectangular shack
When a triangle came down—Keerplunk!
"I must go to the hospital,"
Cried the wounded square,
So a passing rolling circle
Picked him up and took him there.

—Shel Silverstein, *A Light in the Attic* (1981)

Bridges, forces, compression, and tension—oh, my! Can this vocabulary be for kindergarten students? You bet! Come along as the "Straw Bridges" STEM (science, technology, engineering, and mathematics) unit unfolds based on the mathematical standards for the geometry concepts of identifying, describing, analyzing, comparing, creating, and composing shapes. Oh, yes. Squares, rectangles, circles, trapezoids, hexagons, cubes, cones, cylinders, and spheres are all found in the kindergarten mathematics standards. Two very creative teachers from Broadmor Elementary School in Tempe, Arizona—kindergarten teacher Lori Schmidt and fifth-grade partnering teacher Joshua (Josh) Porter—have developed and cotaught the following Straw Bridges STEM unit. The fifth-grade students become "learning buddies" to the kindergarteners and help them with their engineering projects.

At the beginning of any new curriculum planning adventure there is always brainstorming and reflection, and as you will read, this was not a canned STEM unit. Yes, Lori and Josh both had ideas about straw-building activities they had studied, but they took those ideas and developed their own integrated straw bridge STEM unit. Was it perfect the first time through? No, of course not. But like all great teachers, they monitored and adjusted their teaching to fit the students' needs. They also took notes about how to change or enhance the lessons to improve them for subsequent years.

Overview of the Straw Bridges STEM Unit

In Lori and Josh's words

We start the unit planning with a big idea or engineering task. We look at what we want students to have as an end product. In this unit, we started with the kindergarten math standards as the "driver." From there we identified the engineering properties as our "copilot," the technology needed to complete the project. We then decided that the science concepts, while necessary to understand the forces at work, would not be the focal point of the learning and therefore were the "back-seat passengers."

The big idea was to design and build a bridge using students' understandings of various shapes. This was the end product. But the necessary steps to complete the bridge design included allowing the students to tinker with their own ideas for design. Each successive lesson reinforced a concept necessary to build an effective bridge. Each lesson ended with examples of successes and failures in the design phase with reflections to help guide decisions for the next steps in the process.

Once the students have their own ideas for design, we introduce vocabulary as we go. Vocabulary is taught with kinesthetic movements so our English language learners and nonreaders can still demonstrate understanding.

DEVELOPING THE STRAW BRIDGE STEM UNIT

Lori and Josh had already decided on their unit's big idea, which was to have the students design and build a bridge using shapes that included circles, triangles, squares, rectangles, rhombuses, trapezoids, hexagons, cubes, cones, cylinders, and spheres. To get to this transdisciplinary task, they began with the first step in the W.H.E.R.E. planning template, which is to decide on the *what* (see Figure 6.1). What standards will be addressed? What content standards and big ideas will the students need to know and be able to do? What are the desired learning goals for this instruction?

Figure 6.1. The *W* section of the W.H.E.R.E. model

W	**What** are the desired results, including the big ideas, content standards, knowledge, and skills? • List the content standards and what the students will know and be able to do.	**Why** would the students care about this knowledge and these skills? • Craft the driving question that will lead to the development of the integrated tasks that provide for the application of the content, knowledge, and skills. • List the essential questions that can be answered as a result of the learning.

Having selected the *what* items in the first step, Lori and Josh began to think about the second part of the *W* guidepost—the *why*. Why would a student care about learning these concepts and skills? This can be hard to articulate beyond the response "Because it's in the standards." Being able to describe the *why* can help more clearly define the learning outcome. The *why* sets the direction of the instruction and helps craft the driving question that ties together the development of the integrated tasks to provide for the application of the content and skills. Without having a *why*, the learning can become directionless, getting mired in activities without having any real purpose or focus for the student. And while these activities may be fun and engaging experiences for children, they do not aid them in developing an understanding of the core ideas listed in the *what*.

In this example unit, the partnering teachers are from different grades. One of the overall goals is for the fifth graders to help the younger students with their constructions and learn how to cooperate with others. For the younger students, working with the older students gives them more in-depth talking time, as they need to describe their ideas and construction plans to their learning buddies. Various research studies support the positive effects that the role of mentor-mentee has on learning. Findings across many subjects have identified learning as an active process that is highly social and is enhanced by the intentional support provided by more knowledgeable individuals, whether they are peers, mentors, teachers, or experts in the field.

For the Straw Bridges STEM unit, the main drivers in its development were the kindergarten geometry standards from the Arizona State Standards (shown in Figure 6.2, p. 50).

Figure 6.2. The geometry standards for the Straw Bridges STEM unit

Geometry		
K.G.A. Identify and describe shapes	K.G.A.1	Describe objects in the environment using names of shapes, and describe the relative positions of these objects using terms such as *above, below, in front of,* and *next to.*
	K.G.A.2	Correctly name shapes regardless of their orientation or overall size (circle, triangle, square, rectangle, rhombus, trapezoid, hexagon, cube, cone, cylinder, and sphere).
	K.G.A.3	Identify shapes as two-dimensional (lying in a plane, flat) or three-dimensional (solid).
K.G.B. Analyze, compare, create, and compose shapes	K.G.B.4	Analyze and compare two-dimensional and three-dimensional shapes, in different sizes and orientation, using informal language to describe their similarities, differences, parts (e.g., number of sides and vertices/corners), and other attributes (e.g., having sides of equal length).
	K.G.B.5	Model shapes in the world by building shapes from components (e.g., use sticks and clay balls) and drawing shapes.
	K.G.B.6	Use simple shapes to form composite shapes. For example, "Can you join these two triangles with full sides touching to make a rectangle?"

Source: Arizona Department of Education (2016).

In their description of the planning for this unit, Lori and Josh mentioned using the math standards as the "driver." What did they mean? For every STEM unit there may be one discipline—or at the most two disciplines—driving the content standards that are the focus of instruction. In an integrated unit, the central focus for the learning cannot be all the standards covered. Too many different standards being introduced at one time can only serve to create confusion for the learners and interrupt their ability to concentrate on the acquisition of new knowledge and skills. As described in research cited in *STEM Integration in K–12 Education,*

> *Split attention—simultaneously dividing one's attention between competing sources of information—is cognitively demanding and can be a major obstacle to understanding and learning. The split-attention effect is evidenced by difficulties in storing and processing information that is physically separated.* (NRC 2014, p. 84)

Supporting standards can be woven into the learning experiences to provide opportunities for reinforcement and for extended practice of previously learned concepts. To help with this, we use the analogy of a car, first described in Chapter 5. In a four-passenger automobile, there is only one driver—the set of core standards that the instruction will focus on delivering. These cover the primary objectives for the unit and drive the learning forward. In your car, the front-seat passenger represents those standards that may be closely aligned to the driver and are helpful in navigating the instructional route. The front-seat passenger acts as the copilot to the core standards and influences the learning path necessary for greater student understanding. This can help provide context for the learning or make the experiences more rigorous and meaningful.

There may be passengers in the back seat—these are the tangential learning standards that could be tied to individual activities or the culminating product of the unit. For example, in the straw bridge unit, students may use a science text or leveled science readers for content acquisition, which can reflect English language arts (ELA) standards for analyzing informational text. The students will be practicing those ELA skills, but the primary instruction in the unit is not focused on the teaching of reading informational text. Or, students might connect their bridge project to issues that affect their community (whether it is the classroom, the school, or the broader neighborhood), which can tie to social studies standards, but the understanding of "communities" is not the focal point of the learning in the project. These are just tangential standards that help deepen the integrated learning experience beyond the STEM topics. These standards can be opportunities to reinforce previously taught concepts or a way to introduce an idea that will be explored at a future time.

The geometry standards are the driver or primary focus of instruction for the straw bridge unit. But as we all know, it is difficult to separate and focus on only a few standards. Therefore, the beauty of this unit is that it incorporates other standards from both science and ELA. Once you identify the main content driver that you want to stress and really focus the learning around, there also will be other supporting standards (passengers) to consider. This is part of your decision-making process as you begin to think about your STEM unit scenario and start to develop the W.H.E.R.E. template. Integrated STEM units are just that: integrated with more than one standard to help reach the answer to the unit's driving question and to facilitate the completion of the transdisciplinary task.

BRAINSTORMING THE STRAW BRIDGE STEM UNIT

Lori and Josh also realized that they could incorporate standard K-PS2-1, Motion and Stability: Forces and Interactions (see Figure 6.3, p. 52), from the *Next Generation Science Standards* (*NGSS*; NGSS Lead States 2013). This standard was comprehensive enough for them to develop their ideas of having the students begin to understand pushes and pulls as they examined different shapes found in bridges. The students were introduced to local bridge builders who used sophisticated terms such as *compression*, *tension*, *structure*, *support*, and *trusses* in describing the images they shared of how they build bridges. Students' use of this vocabulary evolved naturally as they tried to articulate their understandings of why some shapes were more successful than others. All of this

Figure 6.3. *NGSS* performance expectation for K-PS2-1

<table>
<tr><td colspan="3">K-PS2-1 Motion and Stability: Forces and Interactions</td></tr>
<tr><td colspan="3">Students who demonstrate understanding can:
K-PS2-1. Plan and conduct an investigation to compare the effects of different strengths or different directions of pushes and pulls on the motion of an object.
[Clarification Statement: Examples of pushes or pulls could include a string attached to an object being pulled, a person pushing an object, a person stopping a rolling ball, and two objects colliding and pushing on each other.]
[Assessment Boundary: Assessment is limited to different relative strengths or different directions, but not both at the same time. Assessment does not include non-contact pushes or pulls such as those produced by magnets.]</td></tr>
<tr><td colspan="3">The performance expectation above was developed using the following elements from A Framework for K–12 Science Education (NRC 2012):</td></tr>
<tr><th>SCIENCE AND ENGINEERING PRACTICES</th><th>DISCIPLINARY CORE IDEAS</th><th>CROSSCUTTING CONCEPTS</th></tr>
<tr><td>Planning and Carrying Out Investigations
Planning and carrying out investigations to answer questions or test solutions to problems in K–2 builds on prior experiences and progresses to simple investigations, based on fair tests, which provide data to support explanations or design solutions.
• With guidance, plan and conduct an investigation in collaboration with peers.
- - - - - - - - - - - - - - - -
Connections to the Nature of Science
Scientific Investigations Use a Variety of Methods
• Scientists use different ways to study the world.</td><td>PS2.A: Forces and Motion
• Pushes and pulls can have different strengths and directions.
• Pushing or pulling on an object can change the speed or direction of its motion and can start or stop it.
PS2.B: Types of Interactions
• When objects touch or collide, they push on one another and can change motion.
PS3.C: Relationship Between Energy and Forces
• A bigger push or pull makes things speed up or slow down more quickly. (secondary)</td><td>Cause and Effect
• Simple tests can be designed to gather evidence to support or refute student ideas about causes.</td></tr>
</table>

Source: NGSS Lead States (2013).

led up to the development of their transdisciplinary engineering design task, which was to have the student groups build a bridge that could hold at least five or more books without collapsing.

In the primary grades, we find it difficult to separate and focus on only one of the engineering design standards. It is not that the individual learning objectives of the standards cannot be isolated, but it is difficult to cut short the learning experience when student excitement is at a peak. What fun is it to talk about a problem or generate a solution without the opportunity to try building it? How do you foster persistence, experimentation, and self-reflection if you don't allow time for solution improvements? For these reasons, in this primary-grade unit the supporting content includes all of the engineering design standards (ETS) in addition to the science. There are of course other content standards, but in this unit, the understandings about geometric shapes will be held at the forefront of the planning and will act as the driver. The engineering design standards will play the role of the front-seat passenger.

Math is the main driver for this STEM unit, but as you can see, the engineering design standards and science are essential secondary or supporting standards (Figure 6.3). These are the passengers in the STEM car. As Lori and Josh were designing this STEM unit, part of their decision-making process was to think about the unit holistically, and looking at the standards that needed to be taught and those that could be used to reinforce previously covered standards helped with the development. As the overall idea for the STEM unit takes shape, it then becomes a process of thinking about what interesting or relevant scenario can be crafted, which will lead to the development of the driving question. In this case, the students were learning about different types of bridges, they were using their straws to construct different geometric shapes, they were learning a rich variety of core ideas and new vocabulary words, and they were building comfort in collaborating and communicating with each other (see Figure 6.4 and Figure 6.5, p. 54).

Figure 6.4. Students creating their shape pieces prior to bridge construction

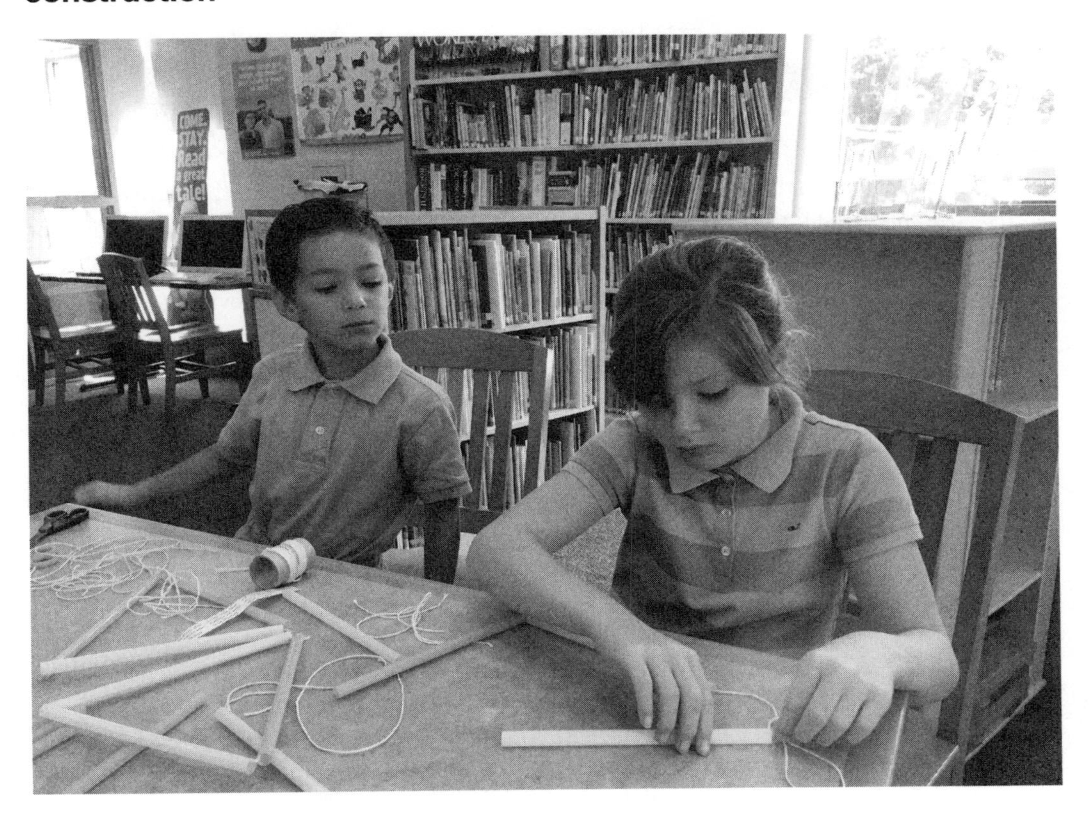

Figure 6.5. Students beginning construction of the straw bridge

Having decided upon the *what* and the *why*, it was now time to put it all together with an engaging scenario that would draw from the essential questions and bring together all the content for the students. This scenario or storyline helps the students become active participants in the learning, while the driving question communicates the purpose for learning by providing a real-world context that the students can relate to. This scenario gives students the opportunity to apply their key understandings in their transdisciplinary task.

THE SCENARIO

In developing this Straw Bridges STEM unit, the teachers aimed to give students experiences with seeing shapes, creating models, drawing shapes, and composing composite shapes found in the real world (math standards K.G.B.5 and K.G.B.6). The students described the shapes by talking about those found in different types of bridges. From there, Lori and Josh decided to bring this home with the following scenario:

> *There was a very heavy rainstorm, and it washed out the bridge to the town. The families needed to return to their homes. They loaded up trucks with supplies but*

they needed a bridge to cross the river in order to get home. You are part of an engineering design team that has been asked to construct a new, strong bridge. But first your design team will need to build and test models to decide on the best and strongest design for this new bridge.

Now, with a scenario in mind and the content standards in place, it was time to create the driving question that would tie together the development of the integrated tasks: "How can we design and build a bridge out of straws with tension and compression so that the bridge does not collapse when tested?" (see Figure 6.6).

Figure 6.6. Students testing their straw bridge construction

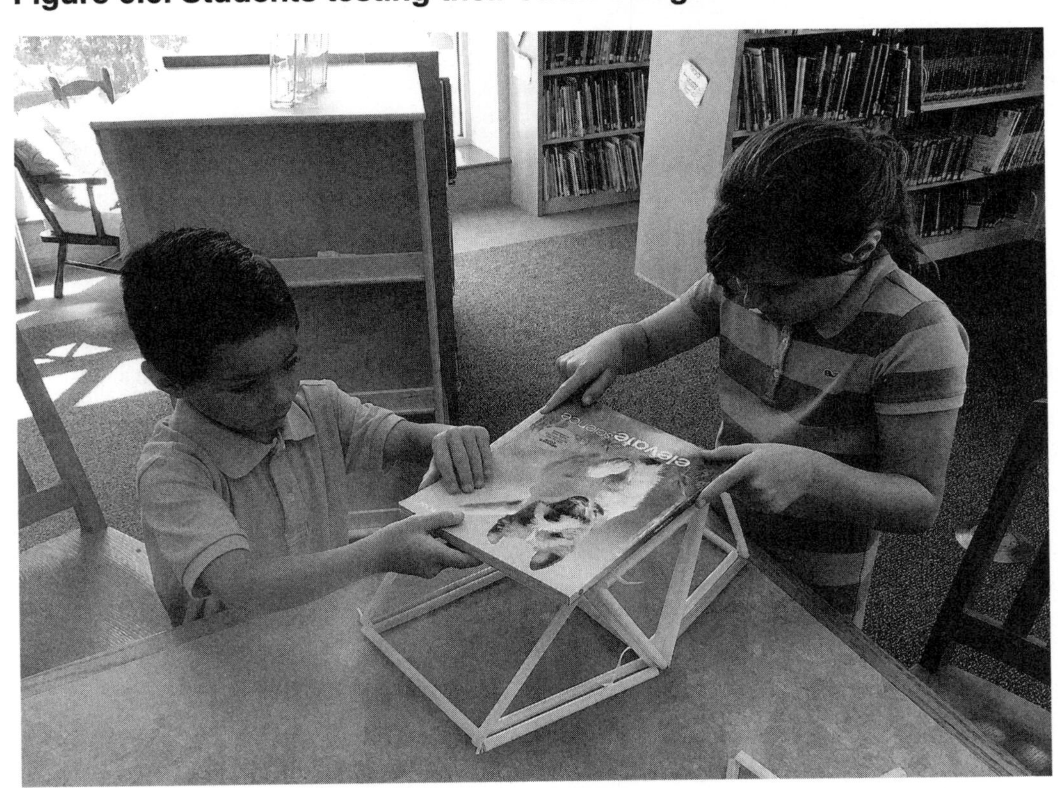

The completed *W* section of the W.H.E.R.E. model (see Figure 6.7, p. 56) provides a description of the planning for the *what* and the *why*. It describes the key drivers for the instruction in the unit.

Figure 6.7. *W* section of the W.H.E.R.E. template for the Straw Bridges unit

<table>
<tr><td></td><td>What
The students will develop an understanding of different geometric shapes through hands-on experiences where they use these shapes to design and build a bridge.

Main standard (drivers)
K.G.A. Identify and describe shapes.
1. Describe objects in the environment using names of shapes, and describe their position using terms such as above, below, besides, in front of, behind, and next to.
2. Correctly name shapes, regardless of their orientation or overall size (e.g., circle, rhombus, trapezoid, hexagon, cube, cone, cylinder, and sphere).

Secondary standards (passengers)
Engineering Design
ETS1.A. Defining and Delimiting Engineering Problems
ETS1.B. Developing Possible Solutions
ETS1.C. Optimizing the Design Solution

Science
Motion and Stability
K-PS2-1. Plan and conduct an investigation to compare the effects of different directions of pushes and pulls on the motion of an object.

English Language Arts
K.RL.1. With prompting and support, ask and answer questions about key details in a text.
K.SL.5. Add drawings or other visual displays to descriptions as desired to provide additional detail.
K.SL.6. Speak audibly and express thoughts, feelings, and ideas clearly.</td></tr>
<tr><td></td><td>Why
Students can better understand the attributes of different shapes when they are used to solve a problem. The bridge problem offers a variety of successful design solutions.

Driving question
How can we design and build a bridge out of straws with tension and compression (push and pull) so that the bridge does not collapse when tested?

Essential questions
Mathematics
• What geometric shapes can be used to construct an effective bridge?
• What total geometric design used to construct our bridge is most effective?
• What are the attributes of the different shapes that were used to construct the bridge?</td></tr>
</table>

Continued

Figure 6.7 *(continued)*

<table>
<tr>
<td></td>
<td>

Essential questions *(continued)*

Engineering Design

- What observations and questions need to be answered before you can solve the problem?
- How can the engineering design process be used to find a solution to your design problem?
- How will drawing a picture or creating a blueprint of the design help with the construction of the bridge?
- Is there more than one design that will provide the best solution for the strongest bridge structure?

Science

- How can we best describe how much force we can place on our bridge?
- What simple tests can be used to determine the stability of the bridge?

English Language Arts

- Using the design of the bridge, how can you explain, in your own words, the two types of forces—compression and tension (push and pull)?

</td>
</tr>
</table>

Special Thanks

A special thanks to Broadmor Elementary teachers Lori K. Schmidt and Joshua Porter for opening up their classrooms and sharing the Straw Bridges STEM unit. Following is a bit more about these two dynamite teachers and how they became interested in implementing STEM instruction in their classrooms.

In Lori's words

How did I become interested in STEM? My daughter is a senior at Northeastern University and majoring in mechanical engineering. My interest in STEM started with my daughter being in a field dominated by males. It is very important to expose all children to STEM, but getting girls excited about these particular fields has been a focus of mine since I started teaching in the 1980s. If I can give my students opportunities to explore STEM and develop a passion in this area, then hopefully they will seek out jobs in a STEM field.

In kindergarten, the vocabulary learned, the critical thinking processes used, and the development of fine motor skills are enhanced greatly by using STEM. In order to compete in this global world/economy,

Continued

(*continued*)

we need to start teaching STEM when the children are very young, and that is why I started in kindergarten. STEM is fun for me to teach and for my students to learn.

In Josh's words

I have been teaching in the Tempe School District for 10 years and teaching in some capacity for 17 years. I believe in teaching to the head, the hands, and the heart to help develop well-rounded citizens. I focus on integrating English language arts, technology, kinesthetic learning, art, and STEM to create a diverse and innovative classroom. I think that teaching techniques that are a benefit to some will benefit all.

I first began teaching in a STEM-oriented way because conventional math and science classes were not focusing on science and engineering concepts. My goal is to develop STEM units based on real-life scenarios that give children context for the importance of science, technology, engineering, and math. The key to collaboration comes from children of different ages discussing, inquiring, designing, experimenting, and reflecting. I believe that learning by itself isn't fun, but that "fun" is the byproduct of high-quality learning experiences!

REFERENCES

Arizona Department of Education. 2016. *Arizona mathematics standards—Grade K.* Phoenix: Arizona Department of Education.

National Research Council (NRC). 2012. *A framework for K–12 science education: Practices, crosscutting concepts, and core ideas.* Washington, DC: National Academies Press.

National Research Council (NRC). 2014. *STEM integration in K–12 education: Status, prospects, and an agenda for research.* Washington, DC: National Academies Press.

NGSS Lead States. 2013. *Next Generation Science Standards: For states, by states.* Washington, DC: National Academies Press. *www.nextgenscience.org/next-generation-science-standards.*

Silverstein, S. 1981. *A light in the attic.* New York: Harper & Row.

CHAPTER 7

Developing a STEM Unit With Engineering as the Driver—Baby Bear's Chair

Once upon a time there were three bears who lived in a house in the forest.

There was a father bear, a mother bear, and a baby bear.

—British 19th-century fairy tale

Creating engineering tasks based on literature is nothing new in primary grades. Fairy tales are usually a class favorite because the stories allow young children to get lost in the tales and fantasy. The students enjoy the characters and the problems the characters encounter, and even more intriguing is the opportunity for them to possibly design a solution! Fairy tales are a wonderful way to use literature to help engage students in meaningful engineering design challenges and authentic opportunities for problem solving and critical thinking.

Many of the new science standards emphasize having students engage in the practices of scientists as well as engineers. However, there is scant research on how these engineering experiences help young students, especially in the areas of developing language fluency and literacy. Most of the work focuses on middle- and secondary-level students. Generally, these types of engineering tasks require sustained attention over longer periods of time, greater cognitive concentration, and a level of independence to achieve task success. However, recent studies have begun to show that integrating age-appropriate engineering-type readings in the K–2 classroom can provide students with access to real-world examples and the application of problem-solving skills. Marley and Szabo (2010) documented that in the kindergarten context, narrative texts, active experiences, and producing simple written texts help students engage in further discourse as they build on their ideas, expand their understanding of core concepts, and use appropriate vocabulary.

On the following pages, you'll see how a kindergarten teacher uses literature as the context for her students to engage in the engineering design process. This process is one that guides engineers as they create solutions to problems presented to them. It is a series of steps that emphasizes open-ended problem solving and takes into account the criteria needed to define a successful solution; it also considers the limits or constraints around which the solution is required. There are several different versions for the number of steps in the process, but they all focus on the following: (1) defining the problem, (2) proposing a solution, and (3) testing and improving the solution.

Allison Davis is an incredible kindergarten teacher in Arizona who is very intuitive about what her young students know, and she honors how those streams of knowledge contribute to the learning environment in her classroom. Whether she is teaching second-language learners who are often experiencing formal education for the first time upon entering her classroom, or kindergartners who have already been identified as academically talented or gifted and are very eager to take on higher-level tasks and challenges, Allison invites them all into her community as equal contributors. It is exhilarating to visit her classroom and watch these young children thrive and enjoy the possibilities that each lesson presents.

The STEM (science, technology, engineering, and mathematics) unit that Allison shares in this chapter not only supports engineering, but is an excellent example of how the integration of math and literature into the unit also supports a meaningful learning experience. This experience provides students with the opportunity to build conceptual understanding, strengthen the foundational skills necessary to become fluent readers and writers, and apply what they are learning in an authentic way. Throughout the unit, Allison focuses on allowing her students time to apply what they are thinking, to make sense of the new knowledge, and to work through things that do not yet have meaning for them.

Figure 7.1. A chair holding Baby Bear with the reference of being three cubes high

ALLISON'S STEM JOURNEY, TOLD IN HER WORDS

One of my favorite engineering-through-literacy units is "Baby Bear's Chair." This engineering task was inspired by the book *A Chair for Baby Bear* by Kaye Umansky (2004). The story takes place a few days after Goldilocks broke Baby Bear's chair. Papa Bear has not fixed the chair yet, so the bears go shopping for a new one. Along the way Baby Bear dreams of all the different types of chairs he might get. But in the end, none of the chairs are just right. Coming back home, Baby Bear sees a box at his front door. It is from Goldilocks.

I use this first part of the book to get students engaged and interested, and then I present them with an engineering task to create a chair for Baby Bear that is just right. I provide each group of students with four index cards to build the chair. Of the index cards they can choose from, one card is big (5 × 7 in.), two are medium sized (4 × 6), and three are small (3 × 5). The build criteria state that the chair must be three linking cubes off the ground, be able to stand by itself, and be able to hold Baby Bear for at least 15 seconds (see Figure 7.1). Besides the materials, I also place constraints on the time (30 minutes to build) and the group size (three people per group). During this unit, we also learn about two-dimensional (2-D) and three-dimensional (3-D) shapes in math. I feel that to understand shapes, you have to apply the new knowledge to an engineering task.

The criteria, materials, and constraints have been a work in progress. The first year I did this build, I let the kids use four of any size index card. Every group picked four big cards. Their chairs, as well as materials used, were very similar. Putting constraints on the size of index cards caused students to make decisions about what would be the best to create all parts of their chair. The next year, my class almost outsmarted me by presenting the idea that to create the chair they could fold the index card once and have it sit on the floor. They talked about the chair they use in their playroom when they watch TV and play video games. They were not wrong—that was a type of chair—but to use the 3-D shapes I had envisioned, the challenge needed a chair that was off the ground. Being that the students were in kindergarten, they were very happy to let me amend the criteria to say that the chair must be three linking cubes off the ground.

This build comes at the end of our math unit. Students have worked with different materials to build and study circles, triangles, rectangles, triangular prisms, rectangular prisms, and cylinders. I have created a list of material supplies that are readily available in my classroom, and I introduce them in a particular sequence and then use them in challenges. Index cards are the first materials I introduce and plan challenges around. By the time they get to build Baby Bear's chair, the students have had multiple opportunities to roll, fold, cut, and tape index cards. Using a material that is known helps students concentrate on creating a chair that is unique yet meets the criteria.

To begin the build, the students brainstorm ideas individually. I purposefully do not put them into groups until they have time to get their own ideas down on paper. I challenge students to get at least two ideas prepared. The paper they use is divided into four squares. At this point in the year students have worked on creating diagrams with labels. We have also talked about "faucet" and "funnel" thinking. I encourage them to "faucet think," that is, to get all their ideas down on paper without taking time to evaluate them. Like an open faucet, the ideas should just flow from their minds to the paper. I don't want them to evaluate, or "funnel," their ideas until later.

After this initial brainstorming session, I have the students begin to analyze their initial ideas. I use the strategy that Edward de Bono shared in his 1985 book, *Six Thinking Hats*, in which different color hats represent different thought perspectives. I ask the students to put on their "red" thinking hat and decide which design they love or like the most. Next, they use their "yellow" thinking hat to tell of an advantage that their favorite design would offer Baby Bear. Then, the students are asked to put on their "black" thinking hat and write down something about their design that worries them (see Figure 7.2, p. 62). This helps them reflect on their designs and think through aspects they might not have considered in their brainstorming session. Now it is time for them to meet with their group.

Figure 7.2. Student "blueprint" showing tape, cards, and the worry that it will fall over

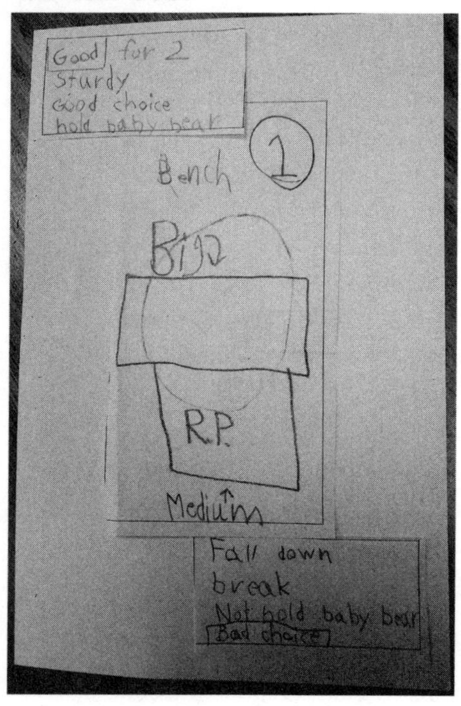

I put the students in groups of three. Using the Six Thinking Hats strategy has increased group discussions during this step. Before I started with this method, students would show their partners their designs and have little to say. Now they use the thinking hats as a way to tell each other which design is their favorite, what they like about the design, and what makes them nervous. Being able not only to talk about the good ideas, but also to identify a potential problem, opens them up to listening to their group and collaborating and compromising on the final group idea.

I like to keep the groups to three people when they share their ideas. I found that pairs of two will not compromise. They both have an idea and it is hard to get them to budge. With groups of three, they can put their ideas to a vote and the majority rules. While having to compromise and vote isn't always popular with students, it does allow them to keep moving and know that even though their idea was not selected, some part of their idea might be used when deciding another part of the build.

After the first brainstorm, I have noticed that the students' idea of a chair can be very narrow. This makes sense because as they are brainstorming, they are sitting on a chair. Most students think of a rectangle with four legs. To broaden their ideas of what a chair is or could be, I have everybody come back together as a whole group and brainstorm about all the kinds of chairs they have seen or may have in their homes. We also refer to *A Chair for Baby Bear* to see that chairs come in many different shapes, numbers of legs, and purposes. This helps students think beyond the four-legged variety.

There was one class in particular that ran with the idea of different chairs. One student thought of a game chair, with cubes on either side of the seating spot. Her reasoning was that she could put her iPad on one cube and her drink on the other, and in case she spilled, it would not upset her mother. (It turns out this was a real-life situation!) Another student thought of a safety chair. In case it broke, or someone fell off, there were bandages in a pocket on the back and a seatbelt. Another student thought that the chair should be fun. He wanted a slide on the back side of the chair. Yet another student wanted to play on the term *high chair* and made her chair the highest in the air. All these students had great brainstorming ideas, but they needed the whole group to buy in to make it work.

After the students have created and shared their final team design, it is time to "go shopping" for their four index cards. When they start to build, I am always faced with the same dilemma—to allow scissors or not. For this build I do not put scissors out on the table, but if students ask to use them, I find out their purpose first before making my decision. I have seen many a good group and a good project derailed by students who cannot stop snipping away at the index card. If groups are not using their materials responsibly (i.e., they are cutting index cards unnecessarily), I will not let them have a new index card.

I once had a group, two boys and one girl, who were working on their final design. The girl thought that the part of the chair to put Baby Bear on should be a circle. The boys disagreed with her—the notecard was already a rectangle. The girl noticed the scissors on the table and decided they could cut the circle. The boys agreed with her and they kept plugging along. When it was time to make their chair, she told the boys to cut a circle. They cut and it wasn't right, so they continued cutting. The girl assured me it would work once they made a perfect circle. On my second check-in with them, they were working with a circle that was smaller than your thumbnail. We had 3 minutes left in the build. I asked them to wait until after we had shared, and we would come up with a solution to their problem.

During sharing time, I asked all groups who had used scissors to raise their hands. Only the one group raised their hands. I asked the other groups why they didn't use scissors. They said that their shapes could be made with a fold, rolls, or tape. The group with their index cards in shreds asked if they could try again and not cut. I realized that the temptation of the scissors might be too great. If scissors are on the table, students might think they have to use them. By taking the scissors out of the equation, I found that the groups were more successful and had to think more carefully about *why* they would need to request scissors.

(*Safety note:* Remind students that scissors can be sharp and can cut or puncture skin, so they should be handled with care. The use of scissors is to be done only under direct adult supervision.)

During the build, students may only test their design once with Baby Bear and the cubes. The groups have found a work-around to this idea. They find items that are comparable in weight and size to the bear and cubes and use those when they want to test their designs. This incorporates nonstandard measurement and shows me that they can use their resources in different ways. When building the chair, groups often run into challenges. I try to question the groups as much as possible to help them build successfully. Besides circles, the shape that gives them the most trouble is the cylinder. They have trouble rolling all of the index cards evenly and attaching them to the chair. I encourage students to look at the charts around the room and think about all the different shapes that could be used.

It is also helpful if their base is only one shape, like a triangular or rectangular prism. The more shapes you have for the base, the more opportunities there are to make a mistake in trying to create that design. I use masking tape in challenges because it is easier for my students' small hands to use, and I find it easier to put on and take off the index cards. We have a lot of conversations about tape and the correct way to use it. Again, if groups are misusing their tape, they cannot have any more. But I do make exceptions for a group that gets tangled up in it, or if it loses its stickiness due to their redesigning efforts.

After the build phase, I find that the reflection phase is the most critical. Over time, I have learned that it is important for all students to see that they were successful, even if their chair design didn't hold up Baby Bear. I stress that they learned something new and that they tried their best. I try to not define a successful build as only one that meets the criteria. While that is part of it, it is not the only part. I ask students to write their reflection with a framing question, "What do you know now that you DID NOT know before?" This question can be about the build or about working together as a group. I have the groups share their ideas in a community circle before we test the chairs. After placing Baby Bear on every chair and measuring it with the three linking cubes, we talk about any commonalities we see in the chairs that met the criteria (see Figure 7.3). We usually find that the chairs had no more than two shapes in their base and used most of their materials.

Figure 7.3. Completed chairs with Baby Bear sitting on them

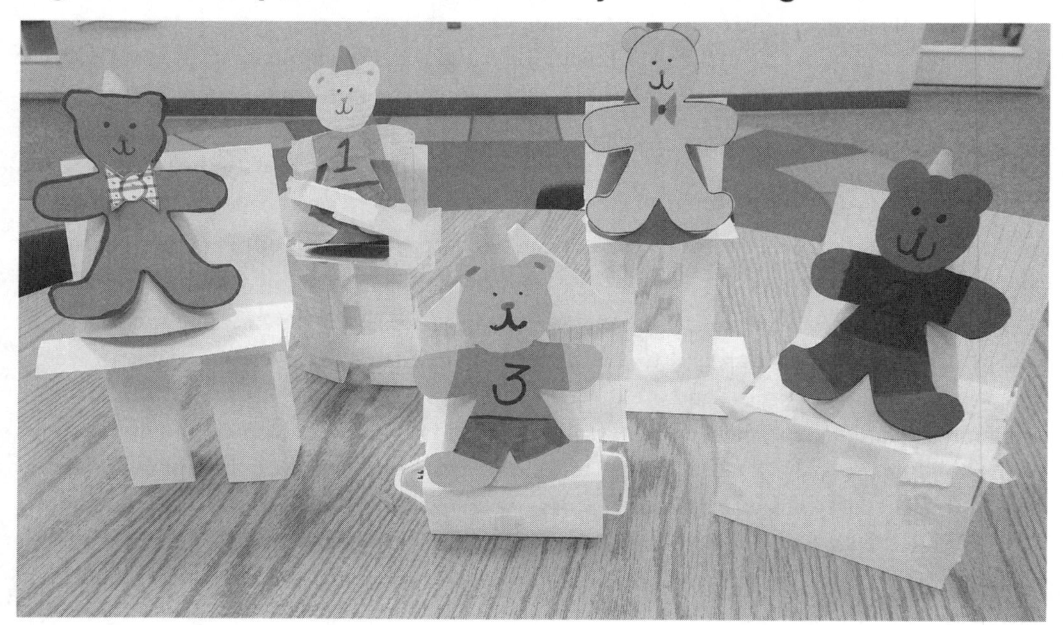

I continue this challenge by posing the following problem: Suppose there is a chair store that wants to produce only one type of chair. It wants to make the *best* chair that it can. Of course, everyone thinks that their chair is the best! We brainstorm what might make a chair "the best." Is it how strong it is? Or how much it would cost? Or the chair that people like the best? Depending on their brainstorm answers, we design a test to find the best chair (see Figure 7.4). To test for the strongest chair, we pile cups of beans on the chair. We record how many groups of tens and extras each chair can hold. Sometimes we say it can get an excellent rating if it holds more than 50 beans. Other times we might pile them on until it breaks. I would recommend doing this test last.

Some years we look at the cost of the chair in tens and ones, or in terms of money. This also requires us to decide what is better, a high-cost or a low-cost chair. The test that my class likes to do the most is seeing what chair that people like the best. To test customer appeal, we set the chairs out in the cafeteria and give one cube to each fifth grader. The students come in and review all the models and then they vote for their favorite chair with their cube. We count the cubes by putting them into piles of tens and ones. We then create graphs and compare the totals.

Figure 7.4. Multiple chair designs for the "best chair" competition

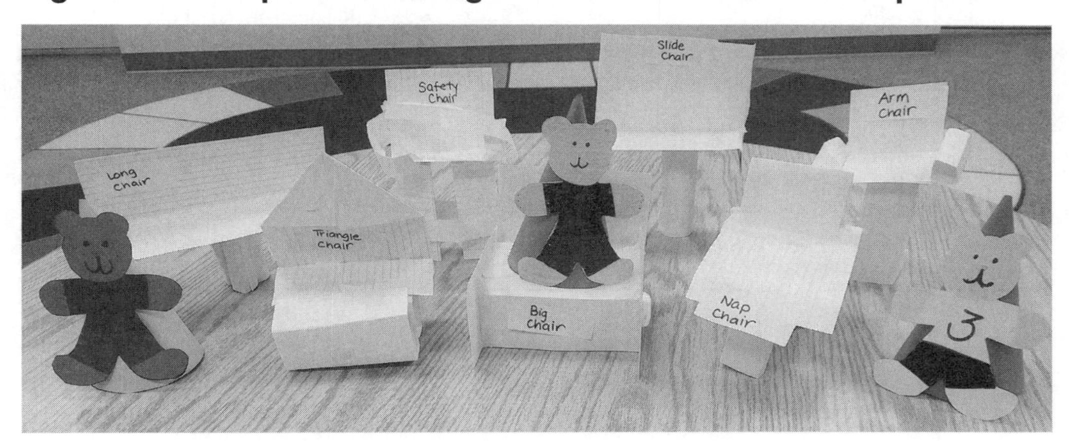

The competition that this creates can sometimes be hard for those who don't win. Throughout the year we talk about how sometimes there is a final winner, but we all win when we learn. We talk about what we might do the next time. In the past, fifth graders have left comments about why they liked each chair. The kindergartners love reading notes from the older kids. All students get a note, and that helps the class with picking the one chair to be featured. The students can also create posters and try to persuade the chair store owner about why their design is the best, even if it didn't win any of the tests for strength or cost, or on its appeal.

This build is one of my favorite engineering tasks. I love how it ties together literature, math, and engineering. It makes the learning fun and includes all students in sense-making and in collaboratively working together. I also love how it teaches my students to be critical thinkers and problem solvers who just might grow up to be engineers!

The completed W.H.E.R.E. model (see Figure 7.5, p. 66) provides a description for each of the planning sections.

Figure 7.5. W.H.E.R.E. planning template for the Baby Bear's Chair unit

<table>
<tr>
<td>W</td>
<td>What are the desired results, including big ideas, content standards, knowledge, and skills?
<ul><li>List the content standards and what the students will know and be able to do.</li></ul>Driver
K-2-ETS1-1. Ask questions, make observations, and gather information about a situation people want to change to define a simple problem that can be solved through the development of a new or improved object or tool.

K-2-ETS1-2. Develop a simple sketch, drawing, or physical model to illustrate how the shape of an object helps it function as needed to solve a given problem.

K-2-ETS1-3. Analyze data from tests of two objects designed to solve the same problem to compare the strengths and weaknesses of how each performs.</td>
<td>Why would the students care about this knowledge and these skills?
<ul><li>Craft the driving question that will lead to the development of the integrated tasks that provide for the application of the content, knowledge, and skills.</li><li>List the essential questions that can be answered as a result of the learning:<ul><li>How can you turn two-dimensional objects into a strong three-dimensional object?</li><li>How do engineers collaborate and communicate to construct objects?</li></ul></li></ul></td>
</tr>
</table>

Continued

Figure 7.5 *(continued)*

How do I plan to meet this goal?

- Identify the pathway, including major tasks and milestones that result in answering the driving question.

Day 1. Read *A Chair for Baby Bear* by Kaye Umansky. While reading aloud, stop after Baby Bear has brainstormed three chairs. What other chairs might be available at the chair store? Students work alone for 5 minutes to "faucet" think about all the different chairs that might be available. They share their brainstorms as a whole class. Continue reading the story. Do not show the final page of the book. Instead, introduce the challenge.

Criteria

- The chair must be three inches off the ground.
- The chair must stand up by itself.
- The chair must hold Baby Bear for at least 15 seconds.

Constraints

- Students work in a group of three to build.
- They have 20 minutes to build.
- Materials: They can use up to four index cards (one can be 5 × 7, two can be 4 × 6, and three can be 3 × 5).

Have students work alone to brainstorm and sketch their designs. Make sure they label their thinking so they can share it on day 2.

Day 2. Students look back at their initial designs. After deciding on their favorite design, they think about what they like about it and what challenges they think it might have. Groups are either self-selected or teacher selected. The group meets, and one at a time, the students share their favorite design. The group then creates its own design. Students can each use a different color pencil and add details to the final sketch and then decide on their shopping list.

Day 3. The group gets 20 minutes to build its design. Once during this time, the group can test its design with the linking cubes and Baby Bear. After 20 minutes, all groups share and test their designs.

Redesigning can take place if the teacher or groups want or need to.

Continued

Figure 7.5 (*continued*)

E	What **Evidence** of learning will be used, and how will I **Evaluate** the final product or project? ***Preassessment.*** What prior knowledge is needed for this task? • Identify the prerequisite skills and understandings. Students must have knowledge of 3-D shapes and how to transform index cards. Evaluation: Students can be evaluated on how well the group works together and shares ideas, materials, and the workload. The students can be evaluated on their design process reflection.	***Formative.*** How will I measure student progress toward understanding? • Establish the assessment tools you will use to monitor progress and inform instruction. Formative check-ins will happen through teacher observation. *Formative #1.* Individual brainstorm check: • Did the student attempt to put his or her idea on paper? • Is the diagram labeled? • Can the students identify what they like about their design and what challenges it might have? *Formative #2.* Group sharing and final design: • Does each member of the group share his or her idea? • Does each student ask and answer questions and participate in the group design? (Evidence can be found by the colors of pencils.) *Formative #3.* Building check: • Are all students participating in the build?	***Summative.*** What criteria are needed for students to demonstrate understanding of the standards, content, and skills? • Create a checklist of criteria for use in a rubric. Building criteria: • Is the structure at least 3 inches off the ground? • Does the structure stand up on its own? • Does the structure hold Baby Bear for at least 15 seconds? • Did the group collaborate to complete the task? • Was the student able to verbalize or write about something new that he or she learned?

Continued

Figure 7.5 (*continued*)

	**Rigor**	**Relevance**
	How can I increase students' cognitive thinking? • Identify tasks that can elevate student thinking, improve inquiry, and increase conceptual understanding. ◆ Before the build, examine 3-D shapes and how 2-D shapes can be transformed by folding and taping together index cards. ◆ Brainstorm chairs and think about how they can be constructed. ◆ Look at chairs around the room, identify shapes you see, and think about how they were made.	Does the learning experience provide for relevant and real-world experiences? • Identify current topics and local issues that can make the tasks more engaging. ◆ How can you "fix" your friend's broken feelings? ◆ How can you say you are sorry and mean it? ◆ How can people work together to create something new? ◆ Why are rules important?

	**Excite**	**Engage**	**Explore**
	What is the hook to excite the learner? • Create the scenario to engage the learner: "Goldilocks needs your help. She wants to create the perfect chair for Baby Bear. She knows that the old chair was 3 inches off the ground and was sturdy enough to hold a bear. Can you help her build a new chair?"	How will students be cognitively engaged throughout the unit? • List the STEM practices that will be used as evidence. ◆ Increase collaboration. ◆ Use failure as a learning tool: What didn't work? How will you redesign your chair? ◆ Increase curiosity about everyday items.	What activities will help students address the driving question? • List questions for students to investigate that will lead them to a deeper understanding of the content and skills. ◆ How do your materials help you decide what you can and cannot build? ◆ How can you work with a group to create a better design? ◆ What makes teamwork work?

Special Thanks

A special thanks to Allison Davis for opening up her kindergarten classroom and sharing the Baby Bear's Chair unit. Following is a bit more about this incredible STEM teacher and how she became interested in implementing three-dimensional STEM teaching and learning with her students.

In Allison's words

Why did I become interested in STEM? I do not know what jobs my kindergartners will have when they grow up, but I do know there are certain skills they will need beyond reading, writing, and arithmetic. They need to know how to work as a member of a team. They need to be able to clearly articulate their ideas with words and in other mediums. They need to be able to construct new ideas and think flexibly about their craft. They must connect their prior knowledge and experiences with new understandings in order to make decisions. They also need to know that kindness is the key to making friends and forming relationships at school, at home, and eventually at work.

All these ideas are a part of STEM in my kindergarten classroom. Students get to be true problem solvers who are actively engaged in learning about science, technology, engineering, and mathematics. Students have to try new things and apply their learning in different ways. Participating in tasks like these engineering builds or creating programs that allow students to "fail forward" builds resilience and fortitude. When students fail forward, they are learning from what went wrong and correcting their mistakes or flaws in reasoning to improve their ideas for the next time. Too often children who are faced with failure give up before ever trying to redesign an idea.

Using STEM in my classroom teaches students that it is OK to take chances and learn something new even if it is not successful.

REFERENCES

de Bono, E. 1985. *Six thinking hats.* Boston: Little Brown and Company.

Marley, S., and Z. Szabo. 2010. Improving children's listening comprehension with a manipulation strategy. *Journal of Educational Research* 103 (4): 227–238.

Umansky, K. 2004. *A chair for baby bear.* Oxford, U.K.: Oxford University Press.

Developing a STEM Unit With Science as the Driver—A Pond Habitat

Over the last two decades, a robust consensus has emerged about the significance of providing quality early education to all children, especially to those from disadvantaged backgrounds. ... An impressive body of research has demonstrated that one of the most powerful influences on an individual's future success in school and life is having sustained opportunities for learning in supportive and enriching environments from infancy through age eight.

—*Creating Learning Environments in the Early Grades That Support Teacher and Student Success*
(Farbman and Novoryta 2016)

W. F. Killip Elementary, a school in Flagstaff, Arizona, has been developing and implementing a standards-based STEM (science, technology, engineering, and mathematics) curriculum for more than five years. Killip has an enrollment of just over 500 students living within a one-mile radius. The school has a 43 percent inward-outward mobility rate (the rate at which individual students either move into or out of the school/district in a given year), and 98 percent of the students are on free or reduced lunch. Going to school means more than just learning academic skills. The school provides a place of care, social interactions, and nourishment. It gives the children opportunities to engage in foundational experiences as well as hands-on experiential learning. Over the past four years, Killip students have been fortunate to take part in a variety of STEM education units, most of which have been developed by the Flagstaff Unified K–12 School District staff.

Both public and privately funded studies have examined the effect that a vibrant STEM education has on our national prominence, whether from a national security perspective, for positioning of worldwide leadership, or for simple economic independence. Yet, the STEM pipeline is a leaky conduit of opportunity for many, especially those in economically disadvantaged settings. This pipeline needs to be filled with greater representations of women and underrepresented minorities. Currently, national trends show little improvement in recognizing the financial and long-term employment benefits that STEM careers can have for these demographics. While different factors may contribute to this, recent studies have shown that early intervention can provide the motivation and vision that can encourage minority and female students to participate in STEM studies.

Research cited in *STEM 2026: A Vision for Innovation in STEM Education* (U.S. Department of Education 2016) supports creating a nurturing environment in which students develop a positive perception of STEM-related learning. Interdisciplinary learning experiences include accessible hands-on activities that allow students to engage in real-world tasks, learn from experts, and be culturally supported, and it's an environment where students' experiences with failure can build an attitude of persistence and fortitude. STEM learning should encourage wonderment and discovery as students pursue interests that match their enthusiasm and their ideas.

"Intentional play activities can support this type of learning experience by providing students with time to explore their uncertainties, construct knowledge from experience, and strengthen relationships," according to *STEM 2026*, a report from the U.S. Department of Education (2016, p. 10). Experiences that invite tinkering, discovery, and risk can serve as incubators for rich learning and build a lifetime of curiosity and inquisitiveness about the world.

The Killip Elementary School team has created these foundational structures that support long-term success. They have thoughtfully analyzed their student data and ways to monitor and model student assessment, setting aside dedicated planning time and creating a framework for accessing local talent. In Chapter 10, the Killip principal describes those structures in greater detail. When Killip's second-grade team decided to develop a STEM-focused pond habitat unit, they reached out to an Arizona Game and Fish ranger, an aquatic ecologist, and a landscaper who installs ponds for a local landscaping contractor. Teachers and schools are often hesitant about contacting community partners for their assistance, but as the Killip team found out, most community partners are more than happy to lend their support. The key here is to have a clear plan developed so that you can carefully explain the goal of your project and how the community partner may be able to help. Let's see how the following Pond Habitat STEM unit, which began as a small project for one grade level, expanded to eventually engage everyone in the school.

It is important to note that the unit was developed over a couple of years and was a work in progress as the team developed it. The team focused on a multidimensional approach, combining an emphasis on the core ideas of interdependent relationships in ecosystems with an understanding of biodiversity. Throughout the unit it is hard to isolate one or two of the science and engineering practices as core, because the students were using them in multiple ways in the individual activities and investigations. The box "Developing the Pond Habitat Unit" is an overview of this STEM unit and how it evolved in six key areas.

Developing the Pond Habitat Unit

As told by the Killip second-grade team

INTEGRATION

When we first developed this unit, we used a *multidisciplinary* approach. We were teaching English language arts skills using the science concept of habitats. We moved the unit over to *interdisciplinary* when we made the endpoint of the unit a research project. This gave the students an endpoint and a purpose for their daily interactions with the text. It wasn't until the 2016–2017 school year that we really were able to move this unit into a true *transdisciplinary* approach. We identified our phenomenon as the declining Spinedace fish population, used that as the vehicle to drive the unit, and then focused our ideas around it. (The Spinedace are a group of native Arizona fish threatened because of water diversion practices.) We were finally able to align the instruction of *all* content areas behind that one common phenomenon for the students. Every text they read, hands-on activity they completed, measurement they took, and presentation they made—it all focused on saving the Spinedace (see Figure 8.1).

Figure 8.1. Introducing the phenomenon—the problem of the Spinedace fish

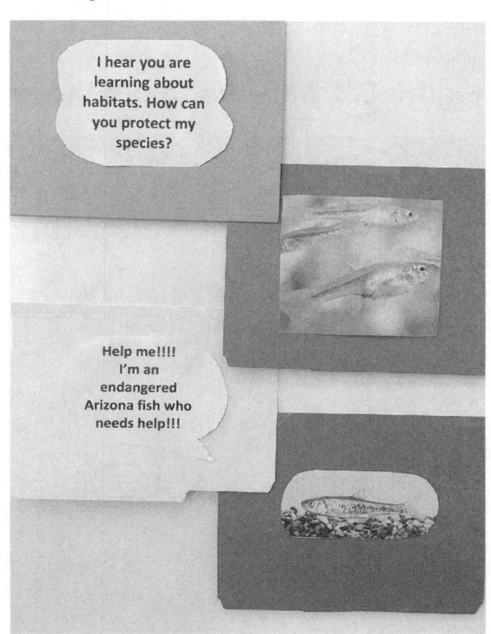

HANDS-ON LEARNING

Initially, the unit included a couple of activities in which the students made a tri-fold pamphlet for a habitat or created an animal for a predefined habitat by selecting body pieces from a set of provided work-sheets. A big step forward in this area was when they started to plan their habitat (ETS1.A: Defining and Delimiting

Continued

(*continued*)

Engineering Problems) and build a model of a habitat out of aluminum lasagna pans and other assorted materials we provided. As you can see in the latest iteration of this unit, the children were doing more than just hands-on activities as they built and labeled their models. They were invested in their work to mark and measure out the plot of land for the pond in the schoolyard and its corresponding components (see Figure 8.2). The students were asked to create two design diagrams (ETS1.B: Developing Possible Solutions) for the pond as they prepared to present their ideas to their peers and to the landscaper. Yes, they had to do the presenting! Part of this strategy, besides keying in on the students' communication skills, was to help them take ownership of the pond habitat itself.

Figure 8.2. Measuring out the area for the students' pond design

TRANSDISCIPLINARY LEARNING AND REAL-WORLD RELEVANCE

Initially, this unit was not project-based or problem-based learning (PBL). The students in previous years proceeded through a series of instructional activities centered on the concept of a habitat, but with no purpose or reason. The content was

Continued

(*continued*)

real-world but not relevant. We entered the realm of PBL, with its increased relevance, when we assigned the research report as the culminating project for the unit. This gave the students a more clearly defined purpose for their learning as they had to obtain, review, and apply their information and skills.

In the most current year, we made the leap into transdisciplinary learning. With the pending extinction of the Spinedace, the real-world relevance of the unit began to drive intrinsic motivation. The students felt pressure and passion to develop a solution to this problem. They knew that if they could, they would be affecting the world in a very real way, far beyond their classroom walls. Student engagement went through the roof, and so did the learning. We saw struggling readers remain engaged with a challenging text because they knew the piece of information they were looking for was somewhere within it. We saw them measure, mark, and label various pond designs so many times that they no longer needed teacher input to find and correct their mistakes. We saw 90 kids sit in the same room with quiet intent as they listened to and discussed feedback on their pond designs (ETS1.C: Optimizing the Design Solution).

ENGINEERING DESIGN PROCESS

Before that year's revisions, the unit did include opportunities for students to go through an engineering design process. However, those opportunities were informal without direct instruction on the steps of the process or on where and how to use it. This time, the opportunities became more intentional and the instruction was provided on numerous occasions. The entire unit was mapped out following an engineering design process, and at several points within the unit, the students worked through the process to arrive at a presentable design.

21ST-CENTURY SKILLS

Too often we put students in groups and tell them to work together to figure it out. Yet, we offer them little or no direct instruction in what that actually looks like. Once again, entering the realm of real-world and relevant PBL allowed us to more thoughtfully plan our instructional activities so that we could focus on direct instruction and application of those 21st-century skills. Students worked in groups in which they had to communicate and collaborate with each other as they created and revised their design plans (see Figure 8.3, p. 76). They then had to evaluate peer designs and discuss strengths and weaknesses in order to arrive at one final design. To secure funding for the needed pond equipment, the students had to develop, practice, and deliver a digital presentation to a local foundation.

Continued

(*continued*)

Figure 8.3. Students show off their completed pond model

COMMUNITY COLLABORATION

In the early revisions of this unit, the only experience we offered our students beyond the classroom was a field trip to a local duck pond. While this allowed them to see what a habitat looked like in real life, it was far from where we have ended up. In this latest year, we still took a field trip to a local pond, but we brought along Mr. Allen, an aquatic ecologist working for a natural resource planning consulting firm. He pointed out the key features of the duck pond that enabled fish, plants, insects, and humans to all interact within a balanced and sustainable system. The students took their knowledge back to the classroom and worked to develop preliminary design plans for a pond at the school.

The students presented their ideas to Mr. William, a water feature contractor with a local landscaping company (see Figure 8.4), and then received his feedback. After a final redesign and Mr. William's approval, the kids developed functional design plans. All they needed

Continued

(*continued*)

Figure 8.4. Students asking questions of the experts

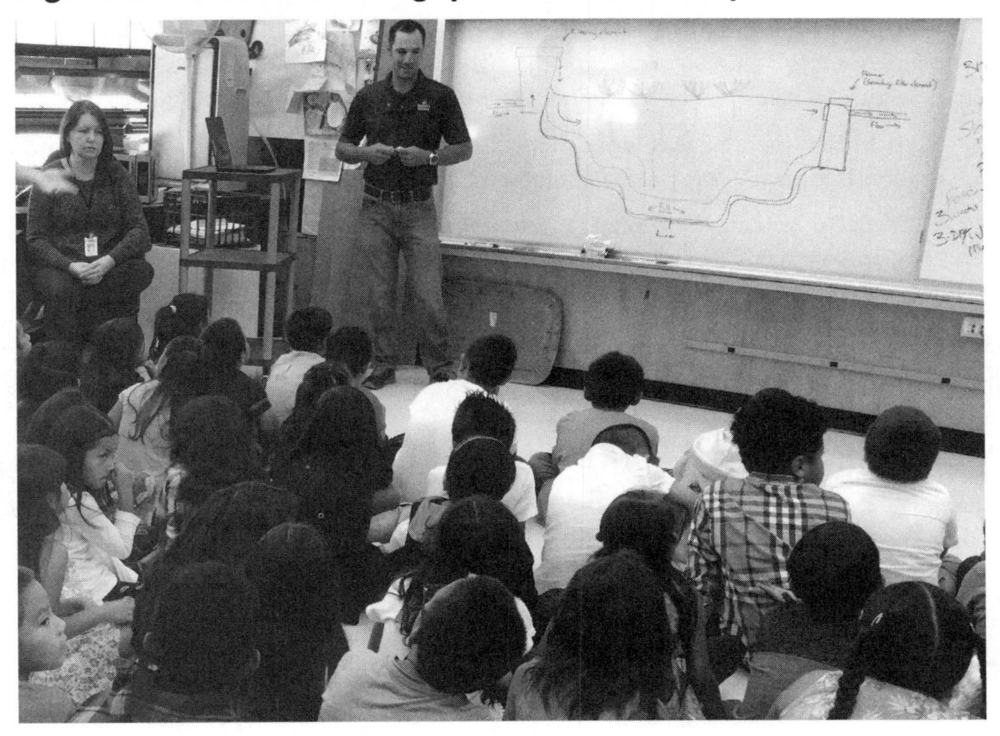

now was money to buy the materials and equipment.

The Forest Highlands Foundation, a charitable organization that promotes community generosity for local improvements and supports the work the Killip School is doing in STEM, agreed to hear the students' funding presentation. On a Friday morning, small teams of students proudly shared their habitat background knowledge, problem, design plans, and budget with the Forest Highlands Foundation representatives. They were awarded funding on the spot, and over our fall break the Killip Park Pond Habitat was installed.

If these are the ways that we want our students functioning beyond their years in the education system, shouldn't it be the way we design learning for them while they are in that system? This notion is ultimately what drives our work in STEM at Killip, and our team is grateful to have made it as far as we have.

The engineering design elements of the *Next Generation Science Standards* (NGSS Lead States 2013) identify three disciplinary core ideas across the K–2 grade span. They are as follows:

- **ETS1.A: Defining and Delimiting Engineering Problems.** Identify a situation that people want to change or create, and through asking questions, making observations, and gathering information about the situation, begin to design a possible solution.
- **ETS1.B: Developing Possible Solutions.** Convey design ideas through sketches, drawings, or models to clearly communicate solutions to others.
- **ETS1.C: Optimizing the Design Solution.** Because more than one solution is possible, compare solution ideas and test designs.

As students work in the Pond Habitat unit, you will see that their activities span all three levels—from defining and delimiting the problem, to developing possible solutions, to finally finding ways to optimize the design.

POND HABITAT STEM CURRICULUM UNIT

Figures 8.5–8.9 (pp. 79–84) show the W, H, E, R, and E, respectively, of the completed W.H.E.R.E. model template for the Pond Habitat STEM unit.

Figure 8.5. *W* section of the W.H.E.R.E. template for the Pond Habitat unit

What are the desired results, including big ideas, content standards, knowledge, and skills?

- List the content standards and what the students will know and be able to do.

Driver

Science—Habitat

2-LS4-1. Make observations of plants and animals to compare the diversity of life in different habitats. There are many different kinds of living things in any area, and they exist in different places on land and in water.

Passengers

Reading

RI.2.1. Ask and answer such questions as who, what, where, when, why, and how to demonstrate understanding of key details in a text.

RI.2.4. Determine the meaning of words and phrases in a text relevant to a grade 2 topic or subject area.

RI.2.7. Explain how specific images (e.g., a diagram showing how a machine works) contribute to and clarify a text.

RL.2.6. Acknowledge differences in the points of view of characters, including by speaking in a different voice for each character when reading dialogue aloud.

RL.2.7. Use information gained from the illustrations and words in a print or digital text to demonstrate understanding of its characters, setting, or plot.

Writing

W.2.2. Write informative/explanatory texts in which they introduce a topic, use facts and definitions to develop points, and provide a concluding statement or section.

W.2.3. Write narratives in which they recount a well-elaborated event or short sequence of events, include details to describe actions, thoughts, and feelings, use temporal words to signal event order, and provide a sense of closure.

Social Studies

2.C1.PO1. Recognize that different types of maps (e.g., political, physical, thematic) serve various purposes.

2.C1.PO2. Interpret political and physical maps using the following elements: alpha numeric grids, title, compass rose–cardinal directions, key (legend), symbols.

2.C1.PO3. Construct a map of a familiar place (e.g., school, home, neighborhood, fictional place) that includes a title, compass rose, symbols, and key (legend).

Math

2.MD.A.1. Measure the length of an object by selecting and using appropriate tools (e.g., ruler, meter stick, yardstick, measuring tape).

2.MD.A.2. Measure the length of an object twice, using different standard length units for the two measurements; describe how the two measurements relate to the size of the unit chosen. Understand that depending on the size of the unit, the number of units for the same length varies.

2.MD.A.3. Estimate lengths using units of inches, feet, centimeters, and meters.

2.MD.A.4. Measure to determine how much longer one object is than another, expressing the length difference in terms of a standard length unit.

Continued

Figure 8.5 (*continued*)

<table>
<tr><td></td><td>

Why would the students care about this knowledge and these skills?

- Craft the driving question that will lead to the development of the integrated tasks that provide for the application of the content, knowledge, and skills. List the essential questions that can be answered as a result of the learning.

Driving question: How can we help save the endangered Colorado Spinedace?

Essential questions
Science

- How do the different elements of a habitat work together to support life?

Reading

- How can I read and use vocabulary words to understand a text about habitats?
- How can I use a diagram to share my ideas?
- How can I identify and use story elements (characters, setting, plot) and voice to better understand a story?

Writing

- How can I create a poem to share information about our pond?
- How can writing help me organize my ideas about different habitats?
- How can I use writing to help me prepare for a presentation?

Social Studies

- How can I create and label a map to help others understand my ideas for our pond?

Math

- How can I use measurement to make sure our pond will be big enough to support the Colorado Spinedace?
- How can I use measurement to make sure we buy all the right-size pond parts and equipment?

</td></tr>
</table>

Figure 8.6. *H* section of the W.H.E.R.E. template

How do I plan to meet this goal?

- Identify the pathway, including major tasks and milestones, that results in answering the driving question.
  - Introduce the problem and brainstorm solutions.
  - Building background: What is a habitat?
    —Text, videos, web searches, journal writing, and graphic organizers.
  - Explore various habitats (ocean, forest, desert, savannah, arctic).
    —Text, videos, web searches, journal writing, and graphic organizers.
    —What do they all have in common?
    —What do those common traits look like for the Colorado Spinedace?
  - Habitat picture activity.
    —What type of habitat do you see?
    —What habitat components do you see?
    —What questions do you still have about this habitat?
  - Freshwater habitats (pond).
    —What does food, air, water, shelter, and space look like in a pond?
    —Text, videos, web searches, journal writing, and graphic organizers.
  - Field trip to a pond with an aquatic ecologist (students take pictures on the trip)
    —Develop a list of criteria for a healthy pond.
  - Create a group pond model.
    —Using aluminum lasagna pans and the materials provided, work with your group to build a model of a pond that could support the Spinedace.
    —Present group models to the class and evaluate against healthy pond criteria.
  - Skype sessions with two staff members at their pond.
    —Staff describe the design and components of their pond that keep it healthy.
  - Professional pond installer Mr. William meets with all classes to review their healthy pond criteria and share additional considerations.
  - Students scout the school grounds for a pond location.
    —Students take and record measurements at the pond location.
    —Students discuss a possible location satellite site for pond components.
  - Students work together in teams to design a pond for the Spinedace.
    —Students present their designs to peers and integrate them into one classroom design.
  - At the same time, students are developing individual brochures for the project.
  - Students present their classroom designs to the professional pond installer and receive final feedback for revisions.
  - One final design is agreed upon.
  - Small groups of students build a slide presentation for potential funders.
  - Students present to a local foundation for funding.
  - Funding is received, and over the fall break the pond is installed.

Figure 8.7. *E* section of the W.H.E.R.E. template

Evidence and **Evaluation**

Preassessment

What prior knowledge is needed for this task?

- Identify the prerequisite skills and understandings:
  - District-level baseline assessments
  - Grade-level, common preassessments
  - Teacher-created preassessments
  - Student writing samples

Formative

How will I measure student progress toward understanding?

- Establish the assessment tools you will use to monitor progress and inform instruction:
  - Various individual, group, and whole-class tools and strategies will measure progress in each content area

Summative

What criteria are needed for students to demonstrate understanding of the standards, content, and skills?

- Create a checklist of criteria for use in a rubric:
  - District-level benchmark assessments
  - Grade-level, common postassessments
  - Teacher-created postassessments
  - Student writing samples

Figure 8.8. ***R*** **section of the W.H.E.R.E. template**

Rigor
How can I increase students' cognitive thinking?

- Identify tasks that can elevate student thinking, improve inquiry, and increase conceptual understanding:
  - Students will acquire information as they read about the components that make up a habitat (food, air, water, shelter, and space).
  - Students will then apply that knowledge to classify various habitats.
  - Assimilating this knowledge, students will develop both a two- and a three-dimensional model of a pond habitat.
  - Finally, students will adapt this knowledge to design, secure funding for, and install a functional pond that can support the Colorado Spinedace.

Relevance
Does the learning experience provide for relevant and real-world experiences?

- Identify current topics and local issues that can make the tasks more engaging:
  - The Colorado Spinedace is an endangered species.
  - It only exists in Arizona's remote streams.
  - How can we be part of the solution?
  - Work with the Arizona Game and Fish Department, take a field trip to a pond with an ecologist landscaping company, and present to potential funders.

Figure 8.9. Final *E* section of the W.H.E.R.E. template

E	**Excite** What is the hook to excite the learner? • Create the scenario to engage the learner: ◆ On the first day of the unit, all second-grade classes will come together, and the teacher will tell the students that he or she is sad. ◆ The teacher will share wonderful childhood camping stories with the students about going camping and playing with small fish in the high-country creeks of Arizona, called the Colorado Spinedace. ◆ The teacher will share that he or she went camping last weekend in the same place, but sadly, there were no Colorado Spinedace anymore. ◆ After doing some research, the teacher learned that the only place in the world that the small fish exists is in Arizona's remote streams and that the population is almost extinct. ◆ The teacher will discuss learning about the habitat, and hopefully together the class can come up with a way to help the Colorado Spinedace. **Engage** How will the students be cognitively engaged throughout the unit? • List the STEM practices that will be used as evidence: ◆ The students will research and compile a list of possible solutions. ◆ They will engage in argument from evidence as they select the best solution (a pond). ◆ They will develop various models for their pond design and select the best according to developed and revised criteria. ◆ They will obtain, evaluate, and communicate information about their project and design to secure funding for the installation of the pond. **Explore** What activities will help students address the driving question? • List questions for the students to investigate that will lead them to a deeper understanding of the content and skills: ◆ What is a habitat? What do various habitats have in common? ◆ What is changing in the Spinedace habitat that is leading to its extinction? ◆ What direct control over the Spinedace's native habitat do we have? ◆ Can we make a habitat for the Spinedace that we have control over? ◆ Who knows more about this than we do, and how can we get them to answer our questions? ◆ How can we develop a presentation that will communicate about our project and inspire potential funders to fund us? ◆ What is the plan for the actual installation of the pond?

Special Thanks

In the Killip team's words

Tremendous thanks and gratitude go to the students and staff at Killip Elementary School who bring this work to life on a daily basis. We are also grateful for the many community partners we have worked with on this and many other projects: Mr. William at Warner's Landscape Company, Mr. Allen at Natural Channel Design, and the Arboretum of Flagstaff. A huge thank you goes to the Forest Highlands Foundation and the Community Foundation of Flagstaff for their financial support of this and many other STEM projects at Killip. We would not be where we are now without all of this support.

Finally, good luck to all of you as you go about developing and implementing STEM teaching and learning in your classroom and school. Just remember—make it real-world and relevant for your students, and they will dive right in!

REFERENCES

Farbman, D. A., and A. Novoryta. 2016. *Creating learning environments in the early grades that support teacher and student success: Profiles of effective practices in three expanded learning time schools.* Boston: National Center on Time & Learning. *https://files.eric.ed.gov/fulltext/ED570690.pdf.*

NGSS Lead States. 2013. *Next Generation Science Standards: For states, by states.* Washington, DC: National Academies Press. *www.nextgenscience.org/next-generation-science-standards.*

U.S. Department of Education. 2016. *STEM 2026: A vision for innovation in STEM education.* Washington, DC: U.S. Department of Education Office of Innovation & Improvement. *https://innovation.ed.gov/files/2016/09/AIR-STEM2026_Report_2016.pdf.*

CHAPTER 9

Moving Students From Inquiry to Application—A Shade Structure

If you tell somebody something, you've forever robbed them of the opportunity to discover it for themselves.

—Curt Gabrielson, author of *Tinkering: Kids Learn by Making Stuff* (2013)

In the Introduction, we saw how a second grader's curiosity about cabbages led to a teacher's inspiring learning opportunity. These are usually referred to as *teachable moments*, but in this case it is a bit of a misnomer. The moment led to an investigation and discussion about cabbages facilitated by the teacher but driven by the students. The students discovered attributes, growing processes, and uses of cabbage by starting with questions they had, and then it was sustained by their drive to learn more about the subject matter. As teachers, we must capitalize on this natural curiosity. The natural inquisitiveness of students can lead to a greater understanding of key concepts, improve their reasoning skills, create disciplinary habits of mind (practices and capacities), and facilitate application of knowledge and skills, thus providing a solid foundation for them as they continue on their lifelong learning journey.

MOVING THE CLASS FROM INQUIRY TO SCENARIOS

In this chapter, contributing author Joel Villegas explains how anchoring your lesson to a real-world phenomenon can help move your students from inquiry to scenarios, or story lines, that they can relate to.

In *STEM 2026: A Vision for Innovation in STEM Education*, the authors state that students who undertake these types of challenges see the relevance of STEM (science, technology, engineering, and mathematics) in their lives and are able to connect it to the benefits it brings to their community: "Undertaking a grand challenge also gives students an accessible entry point as well as the freedom to tinker with ideas because there is no one right answer to solving these issues" (U.S. Department of Education 2016, p. 13). All children, no matter their age, can be presented with phenomena in which they are asked to come up with questions to explore as they work with each other and their teachers as codiscoverers.

The process of intentionally moving students from inquiry to application is not as chaotic as one may think. It may seem that inquiry-based learning is sometimes left to chance and appears too open-ended for practical classroom learning—especially in the K–2 classroom, where foundational skills are usually taught with direct explicit instruction. The need for direct instruction at this level is still prominent, but it can be balanced with inquiry-based and experiential learning as well. With intentional planning, deep questioning, and an arsenal of good teacher moves, the process of guiding students through this continuum of inquiry to application becomes easier than you might expect.

In this chapter, we explain the process while taking you through a first-grade STEM unit that was developed using the W.H.E.R.E. model template outlined in Chapter 5.

ANCHORING TO PHENOMENA

Infants begin to make sense of the world through inquiry as a natural process. This process begins with gathering information and data through the human senses—seeing, hearing, touching, tasting, and smelling. This, at its most basic level, is the definition of science. We learn about the natural world by observing, questioning, gathering data, and reasoning about the results. Effective inquiry is more than just making observations and asking questions. A complex process is involved when individuals attempt to convert information and data into useful knowledge. The useful application of inquiry learning involves several factors: having a context for questions, building a framework around relatable questions, creating a focus for critical questions, and offering different types of questions that address multiple levels of understanding. With thoughtful consideration of the question's purpose, well-designed inquiry learning can produce knowledge formation that may be widely applied.

As Erik M. Francis describes in his 2016 book (*Now That's a Good Question!*),

> *Good questions serve as the formative and summative assessments that measure the extent of a student's learning and they set the instructional focus for an active, student-centered learning experience. ... If your students are demonstrating and communicating—or showing and telling—the depth and extent of what they are learning, then you'll know you've asked a good question.* (Francis 2016, p. 5)

How can we use these questions to ground students' learning in a way that provides the necessary "fertile places" in their minds for them to secure their newly planted understandings as they take root and grow? One way is to use phenomena—rich, relevant, or intriguing experiences that capture students' attention and curiosity while at the same time firing up their thinking.

Developing Anchoring Phenomena

The "anchoring to phenomena" research work done by Reiser et al. (2003) embodies the nature of good inquiry-based instruction. In their work, they refer to these anchoring events as "real-world phenomenological experiences." These experiences provide the anchor, or foundation, that grounds the learning and acts as a tether from which students can connect new learnings. In other words, the anchoring phenomenon becomes the base from which the students' growing understanding will be attached and extended. As these researchers explain,

> *We use anchoring events to help students apply their scientific understandings to the real world. An anchoring event might engage students in observations of their environment or it might also be an initial investigation.* (Reiser et al. 2003, p. 3)

The idea of anchoring to phenomena lays out a plan for helping students develop understanding of subject matter. This requires that teachers know what students already understand and believe about the world. These prior conceptions serve as foundations for building new understandings. Teachers can only use students' prior knowledge if they know what it is. Asking the right kind of context questions can help teachers elicit the prior knowledge that students possess. The purpose for the inquiry is created by using more open-ended framework questions that connect with the authentic interests and curiosities that the students have about their world. Framework questions can scaffold students' thinking and help focus their attention on what they will investigate. The concepts and skills then become the key to unlocking their understanding and help them answer the more focused or essential questions.

The science (or any other discipline subject) is learned through its application to the meaningful problems presented. The students answer the focus question through investigations, data analysis, or the creation of models or some other representation. Through group explanation and argumentation from evidence, a new set of phenomena-driven questions emerges in which students are allowed to refine their thinking through additional exploration. This becomes the sense-making cycle of learning. This process is repeated as often as needed, and the teacher facilitates it until the expected knowledge outcomes and skills are achieved. Figure 9.1 (p. 90) shows a diagram representing this process and is based on the work by Reiser et al. (2003).

With this process as the foundation for good scientific inquiry and learning, what does it look like when applied in a K–2 classroom? How can we get students to take the knowledge acquired at the *disciplinary* level and move it to a *transdisciplinary* level of STEM integration (as defined in Chapter 2)?

Setting Up the Learning

As with any good outcomes-based (standards-based) learning, we begin with the end in mind. What are our desired results? What do we want students to know, and why should they know it? Let's look at a first-grade STEM unit centered around light and shadows. Figure 9.2 (p. 91) shows the *W* section of the W.H.E.R.E. model template.

Figure 9.1. Real-world phenomenological experiences

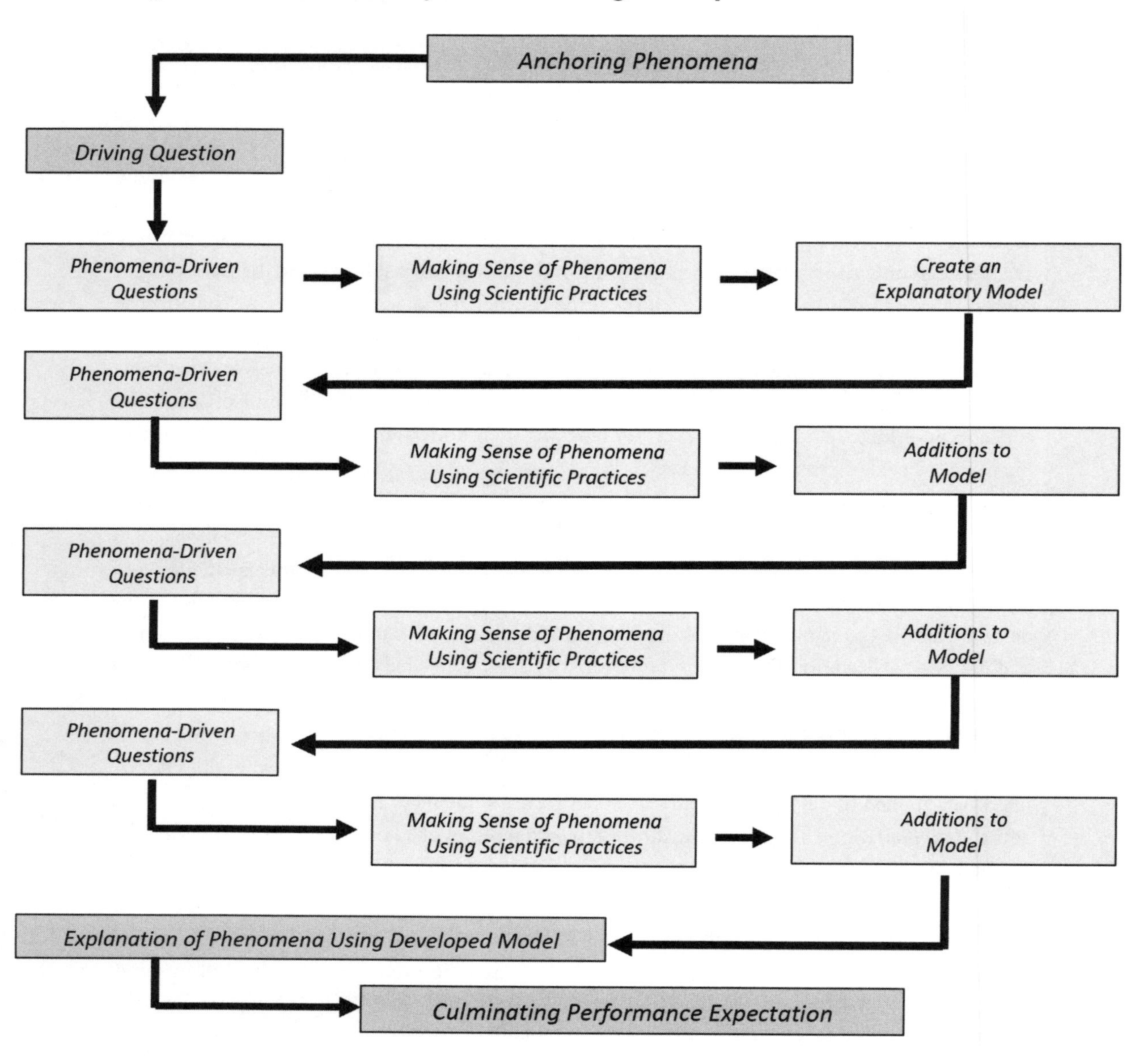

Figure 9.2. *W* section of the W.H.E.R.E. template

W	**What** are the desired results, including big ideas, content standards, knowledge, and skills? • List the content standards and what the students will know and be able to do: 1-PS4-3. Plan and conduct investigations to determine the effect of placing objects made with different materials in the path of a beam of light.	**Why** would the students care about this knowledge and these skills? • Craft the driving question that will lead to the development of the integrated tasks that provide for the application of the content, knowledge, and skills. • List the essential questions that can be answered as a result of the learning: ◆ How does light travel? ◆ Which materials block light? ◆ Why do we need to block light?

In this first-grade example, we wanted our students to be able to accomplish and learn the knowledge, skills, and practices that are associated with the specific science standard of 1-PS4-3, so we made science the driver. The difference between the process we are about to go through and the cabbage example from the Introduction is that we are going to plan for the inquiry rather than have it just arise from an organic situation. If we want our students to learn how light travels, understand the types of materials that can block light or let it through, learn how to plan investigations, and identify cause-and-effect relationships, we must start with a real-world anchoring phenomenon. We must be mindful and purposeful in the selection of the phenomenon to make the learning intentions explicit. Take a minute to think of a demonstration, phenomenon, video clip, picture, story, or other way in which students can begin discussions on these concepts and learning.

You can probably come up with plenty of creative ways to introduce a phenomenon that will anchor the learning. There are books and poems such as *My Shadow* by Robert Louis Stevenson, or online educational videos such as "Shadow: The Dr. Binocs Show," or even interesting shadow images on the internet. We came up with art similar to the photo shown in Figure 9.3 (p. 92) because of the final project the students would be working on.

Students are shown the storybook *The Umbrella Queen* by Shirin Yim Bridges (2008), which contains an image of a woman with a parasol on a sunny day. The students are asked what she is carrying, and most of them answer "an umbrella." A few may refer to it as a "parasol." While there is a difference between an umbrella (protection from rain) and a parasol (protection from the Sun), it is not important that children make this distinction, although it does add to their knowledge base. The students are guided through questioning about the purpose of the parasol or umbrella, and soon they observe that it doesn't look like it is raining. Through the use of open questions, you want to get the students to arrive at the conclusion that the parasol is being used to block the Sun or provide shade.

Figure 9.3. Woman with a type of parasol

After that, it is just a matter of asking scaffolded, strategic questions such as "Why would she want to block the sunlight?" "How does the parasol block light?" and "What materials do you think are good for blocking light?" The teacher can then ask, "How would we know which materials are better for blocking light?" and guide the students through planning an investigation to test their ideas. The students then start making decisions on how to test different materials and determine how to gather data about which materials are best. By having the standard in the back of your mind, as well as what essential questions they will need to address, the learning is then focused on the students being able to do exactly what is stated in the standard.

At this point in the process, the learning is still at the *disciplinary* level. The inquiry has led to our goal of the students learning the science core ideas by engaging in the scientific practices of planning and conducting an investigation along with our infusion of the crosscutting concept of cause and effect. Next, we ask how we can move this to a level of application within the science discipline. Where can blocking light be applied in other areas of science learning?

Applying Within a Discipline

For the previous question, your mind may have raced to engineering (don't worry, we will get there soon enough), but for this unit, the students can apply their newfound knowledge of light, how it travels, and which materials are best for blocking or hindering light to the growth of plants. This new context motivates students to apply their knowledge to other situations. The students can gather data on plant growth by limiting light through materials for plants that may not need direct sunlight. Because growing plants may take too long to gather growth results data within this unit, you can set up for this context by bringing in plant-care cards. These are available online or can be donated by your local nursery or garden center. These cards usually describe care instructions with a label such as "no direct sunlight," "partial sunlight," or "direct sunlight."

Students can test materials by passing light to a stand-in plant and determine which materials meet the plant light category. This is a great way to apply within a discipline. By continuing the discussion of plants needing different amounts of sunlight, you can help your students understand that they can transfer their knowledge of properties of light travel and materials to this new situation. The domino effect will be set in motion. Through questioning and discussion, students can set up new investigations, which will include making observations, gathering data, reasoning about the results, and developing conclusions about light and plants. The original phenomenon about blocking light with a parasol has been applied to the new situation of limiting light for varied plant growth.

Applying Across Disciplines

We can now begin moving students to the highest level of STEM integration, which is *transdisciplinary*. At the K–2 level, this is a very guided and structured process that involves a lot of learning. By definition, transdisciplinary integration involves the seamless application of skills and knowledge while working on a problem or project. For K–2 students, we can give them the same experiences that may appear seamless but are very intentional in their design. Let's begin with an opportunity to brainstorm a scenario that not only will lead to the application of the science concept we started with but can also include content and knowledge from other disciplines. A scenario in this case provides the context for a problem or project that will require students to apply their understandings in a real-world situation. *Scenarios* are situations that provide students with an authentic context, real-world roles, an audience, and the application problem. What authentic scenario can you think of that provides an opportunity for students to apply their newfound knowledge at the transdisciplinary level?

You may come up with a list of ideas for a scenario that could lend itself to investigating which materials would be good for blocking light and how those materials can be applied. Here is the scenario that we used for this unit: Students are shown a video clip from a school nurse on the dangers of high temperatures and heat exhaustion for children on playground equipment in Arizona due to direct sunlight. The students are then presented with the task of designing a shade structure for the playground equipment using various materials and shapes so that it is both functional and appealing.

Because we are adding more content and subjects, we can now revisit and update our W.H.E.R.E. model template (see Figure 9.4) to reflect the new content and add our driving question. This is derived from our scenario that provides the authentic project idea. Notice that we included the concept of shapes in the task and added it to the *W* section of the template. The scenario also helped guide us in selecting the initial image of a woman with a parasol as the anchoring phenomenon, because we wanted the students to consider the purpose of a device to produce shade.

Figure 9.4. Revised *W* section of the W.H.E.R.E. template

<table>
<tr>
<td>W</td>
<td>What are the desired results, including big ideas, content standards, knowledge, and skills?
• List the content standards and what the students will know and be able to do:
1-PS4-3. Plan and conduct investigations to determine the effect of placing objects made with different materials in the path of a beam of light.
K-2-ETS1-1. Ask questions, make observations, and gather information about a situation people want to change to define a simple problem that can be solved through the development of a new or improved object or tool.
1.G.1. Distinguish between defining attributes (e.g., triangles are closed and three-sided) versus non-defining attributes (e.g., color, orientation, overall size); for a wide variety of shapes; build and draw shapes to possess defining attributes.</td>
<td>Why would the students care about this knowledge and these skills?
• Craft the driving question that will lead to the development of the integrated tasks that provide for the application of the content, knowledge, and skills:
◆ How do we design a playground shade structure to protect students from direct sunlight?
• List the essential questions that can be answered as a result of the learning:
◆ How does light travel?
◆ Which materials block light?
◆ Why do we need to block light?
◆ Which shapes block more light?
◆ Which shapes look better?
◆ How do we design and test ideas?</td>
</tr>
</table>

As you can see, the scenario establishes a context for the learning and provides the opportunity for students to apply the science concepts as well as integrating other content areas. The students may have already learned about two-dimensional shapes before this unit, or their use of familiar shapes will provide an introduction to more in-depth shape studies later in another unit. Either way, the integration is purposeful in its intent. Students will have to make and cut out different shapes and test how well they block light. They will learn or apply the engineering design process as they create different structures using different shapes. This is where teachers organize the instruction so there is the illusion of seamless application of knowledge and skills.

Even though this unit has three content areas or disciplines integrated, there is definitely the opportunity to include other content as well. This is always left up to the creativity and instructional needs of the teacher. What other content or subjects would you include in this unit? Maybe you thought of including social studies, geography, or climate. These were a few ideas that jumped out at us as we were developing this unit even further.

SUMMARY

The idea of moving students from inquiry to application can be summarized in Figure 9.5.

Figure 9.5. Moving from the anchoring phenomenon to the final design product

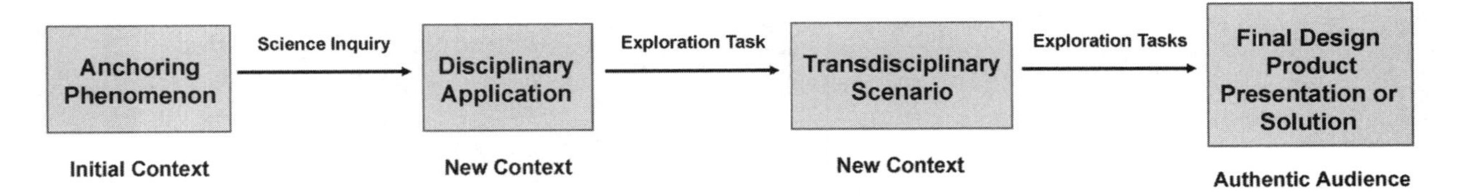

By anchoring to a phenomenon in the real world, students can enter the inquiry at whatever point their prior knowledge and experience allows. The facilitating teacher can then use a variety of inquiry-based methods to get all students to a refined understanding of the scientific concepts. These intentional inquiry methods include strategic questions that move students through the following stages:

1. Concepts about the world (background knowledge, interrelationships, and connections)
2. Data and information (content and disciplinary knowledge)
3. Skills for processing information
4. Application and transfer of knowledge

Questioning is key in moving students through these stages. Exploratory questions, direction questions, checks for understanding questions, and context questions for application are all things that can be used as you move students through this process. The W.H.E.R.E. model template can help you think of these ahead of time and plan how to guide your lesson delivery.

Introducing a new context within the discipline allows students to apply and further make connections to the real world. The scenario allows for the ultimate context that forces students to apply across disciplines in an authentic way. The goal is to get educators to think more broadly of ways to weave the multiple facets of core instruction into a tapestry that showcases the application of experiences that are relevant and real world to the young learner. The question to ask yourself is, "What science inquiry unit do I currently implement that I can take to the transdisciplinary level by using the W.H.E.R.E. model template?" We encourage you to use the W.H.E.R.E. template and try planning your own STEM learning experience.

Figure 9.6 shows the completed W.H.E.R.E. model template for this Shade Structure unit.

Figure 9.6. Completed W.H.E.R.E. template for the Shade Structure unit

<table>
<tr>
<td>W</td>
<td>What are the desired results, including big ideas, content standards, knowledge, and skills?
• List the content standards and what the students will know and be able to do:
1-PS4-3. Plan and conduct investigations to determine the effect of placing objects made with different materials in the path of a beam of light.
K-2-ETS1-1. Ask questions, make observations, and gather information about a situation people want to change to define a simple problem that can be solved through the development of a new or improved object or tool.
1.G.1. Distinguish between defining attributes (e.g., triangles are closed and three-sided) versus non-defining attributes (e.g., color, orientation, overall size); for a wide variety of shapes; build and draw shapes to possess defining attributes.</td>
<td>Why would the students care about this knowledge and these skills?
• Craft the driving question that will lead to the development of the integrated tasks that provide for the application of the content, knowledge, and skills:
◆ How do we design a playground shade structure to protect students from direct sunlight?
• List the essential questions that can be answered as a result of the learning:
◆ How does light travel?
◆ Which materials block light?
◆ Why do we need to block light?
◆ Which shapes block more light?
◆ Which shapes look better?
◆ How do we design and test ideas?</td>
</tr>
<tr>
<td></td>
<td colspan="2">How do I plan to meet this goal?
• Identify the pathway, including major tasks and milestones, that results in answering the driving question:
◆ Session 1. Launch the parasol discussion and questioning (light travel exploration).
◆ Session 2. Plan and carry out the light block activity (analyze data).
◆ Session 3. Discuss the results (assessment on materials that block light; performance-based investigation results).
◆ Session 4. Carry out the plant light exploration and investigation (Part 1).
◆ Session 5. Carry out the plant light investigation and results (Part 2).
◆ Session 6. Launch the shade structure design task with video and discussion.
◆ Session 7. Do the shape construction activity (math).
◆ Session 8. Design drawings of the shade structure using shape and attributes.
◆ Session 9. Do materials testing for the shade structure and determination with evidence.
◆ Session 10. Design the presentations (guided through teacher prompts and questioning).</td>
</tr>
</table>

Continued

Figure 9.6 (*continued*)

<table>
<tr><td rowspan="2">E</td><td colspan="3">Evidence and Evaluation
What evidence of learning will be used and how will I evaluate the final product?</td></tr>
<tr><td>Preassessment. What prior knowledge is needed for this task?
• Identify the prerequisite skills and understandings.
Students need to know that objects in darkness can be seen only when illuminated (K-1-PS4-2).
K.G.5. Model shapes in the world by building shapes from components (e.g., sticks and clay balls) and drawing shapes.</td><td>Formative. How will I measure student progress toward understanding?
• Establish the assessment tools you will use to monitor progress and inform instruction:
◆ Data sheets on light travel
◆ Performance assessment on light investigation
◆ Data sheet on plant investigation
◆ Shape construction assessment
◆ Design of planning booklet</td><td>Summative. What criteria are needed for students to demonstrate understanding of the standards, content, and skills?
• Create a checklist of criteria for use in a rubric:
◆ Knowledge of light travel through various materials
◆ Knowledge of the engineering design process
◆ Knowledge of shapes and the shadows they cast
◆ Presentation and defense of their design</td></tr>
</table>

Continued

Figure 9.6 (*continued*)

	**Rigor**	**Relevance**
	How can I increase the students' cognitive thinking? • Identify tasks that can elevate student thinking, improve inquiry, and increase conceptual understanding: ◆ *Anchoring phenomenon.* Why is the woman carrying an umbrella or parasol on a sunny day? This will lead to a "needs to know" on how light behaves and how materials can block it. ◆ *Materials exploration.* Which materials block light more efficiently? Students must set up an investigation (with teacher guidance) to test various materials and the light they allow to pass. ◆ *Plant investigation.* How can we help block light to plants? Guiding students to the task of limiting light to certain plants allows them to apply the information from previous learning at a higher level of rigor. ◆ *Shade structure design.* How can we design a shade structure to help shade playground equipment effectively? Students will need to understand light at a high level as well as shapes and the amount of area involved.	Does the learning experience provide for relevant and real-world experiences? • Identify current topics and local issues that can make the tasks more engaging. ◆ The students will be solving a real-world problem for Arizona children. Equipment burns on playgrounds as well as heat exhaustion are a severe problem in Arizona schools. Students will apply their knowledge of light travel and materials to develop a solution to this problem.

Continued

Figure 9.6 (*continued*)

	Excite	**Engage**	**Explore**
E	What is the hook to excite the learner? • Create the scenario to engage the learner: ◆ Students will be given the challenge of designing a shade structure for school playground equipment in Arizona. The students can design it by using different shapes based on appeal and coverage and considering different materials for sunlight blocking. They will present their designs to school officials to help make a decision.	How will the students be cognitively engaged throughout the unit? • List the STEM practices that will be used as evidence: ◆ During the umbrella/parasol discussion activity: Students will *ask questions* about light travel. ◆ When testing materials, students will *plan investigations*, *analyze data*, and *construct explanations*. ◆ When developing plant light tests, students will *plan investigations*, *interpret data*, and *engage in argumentation from evidence*. ◆ During the shapes attributes activity, students will *look for and make use of structure* and *attend to precision*. ◆ During the shade structure design, students will *define problems*, *design solutions*, and *engage in argumentation*. They will also *use appropriate tools strategically*.	What activities will help students address the driving question? • List questions for students to investigate that will lead them to a deeper understanding of the content and skills: ◆ *How does light travel?* Explore with flashlights on the directionality of light. Test at different angles. ◆ *Which materials block more light?* Explore materials that block and let through light. ◆ *How does the amount of light affect plants?* Set up an exploration to test the effect of direct, partial, and no direct sunlight on plants. ◆ *What are the attributes of shapes?* Explore edges, roundness, points, corners, etc., of various shapes and what makes them that shape. ◆ *What shapes block more light?* Cut out and explore which shapes cast a bigger shadow (the most area).

Special Thanks

A special thanks goes to contributing author Joel Villegas, an experienced science and STEM educator who wrote the text for this chapter. He currently serves as director of the Pinal County School Office Education Service Agency in Arizona.

Joel has experience as a classroom teacher, a professional development planner and presenter, a grants program manager, and a published coauthor of *STEM Lesson Guideposts* (Vasquez, Comer, and Villegas 2017). For 12 years, he taught elementary students as a classroom teacher. Joel has developed and provided professional development resources and support for STEM teachers and leaders in all Pinal County school districts.

REFERENCES

Bridges, S. Y. 2008. *The umbrella queen.* New York: Greenwillow Books.

Francis, E. M. 2016. *Now that's a good question! How to promote cognitive rigor through classroom questioning.* Alexandria, VA: ASCD.

Gabrielson, C. 2013. *Tinkering: Kids learn by making stuff.* Sebastopol, CA: Maker Media.

Reiser, B. J., J. Krajcik, E. Moje, and R. Marx. 2003. Design strategies for developing science instructional materials. Paper presented at the Annual Meeting of the National Association of Research in Science Teaching, Philadelphia.

U.S. Department of Education. 2016. *STEM 2026: A vision for innovation in STEM education.* Washington, DC: U.S. Department of Education Office of Innovation & Improvement. *https://innovation.ed.gov/files/2016/09/AIR-STEM2026_Report_2016.pdf.*

Vasquez, J. A., M. Comer, and J. Villegas. 2017. *STEM lesson guideposts: Creating STEM lessons for your curriculum.* Portsmouth, NH: Heinemann.

Transforming Into a Successful STEM School

STEM is doing for mainstream math and science what school buses did for leveling educational opportunity in the 1920s—creating access to gateway courses for more students, and making such courses matter again.

—Jeff Weld, *Creating a STEM Culture for Teaching and Learning* (2017)

Creating a successful culture of STEM (science, technology, engineering, and mathematics) teaching and learning in any school requires the commitment of various education partners working together toward a common goal. The daily life of the students within the school is touched by many different individuals. How can these individuals create a more holistic and purposeful experience for all students regardless of their intellectual ability or economic status? How do you create a partnership between the school and the home so that they work together as effectively as possible toward common goals? In what ways can you facilitate a common understanding and benefit of that goal for all students? The ultimate goal is to create a unified vision of what students can achieve to become successful—however that success is defined beyond the classroom.

The research is clear and supports the proverb "It takes a village to raise a child." As described in *STEM Integration in K–12 Education*,

> *Although standards, assessments, and educator expertise must be attended to in implementing integrated STEM, the larger context of the school—its policies, norms, practices, and administrative leadership—is also important. Schools are influenced by the norms, practices, and policies of the school district … and are affected by the parents, taxpayers, higher education, and business leaders in the community. … To understand how these different factors may encourage or discourage effective implementation of integrated STEM experiences, it helps to view the education enterprise as a complex system with interacting parts.* (NRC 2014, pp. 127–128)

In Chapter 2, "Pioneering Into STEM Integration," and Chapter 8, "Developing a STEM Unit With Science as the Driver," we provided examples of STEM units that were developed by teachers from W. F. Killip Elementary School in Flagstaff, Arizona. A key factor in the successful implementation of STEM teaching and learning in Killip was the leadership of the school. Principal Joe Gutierrez made the commitment to facilitate this transformational change—one that would last, and not just become another new initiative that would be gone within a few years.

Principal Gutierrez, working with his staff and other STEM education professionals, based the original implementation plan on four core systems that they felt had to be in place to facilitate and maintain this kind of change. Following is a brief summary of those four core systems. The good news is that the core systems are still in place today and have continued to facilitate STEM teaching and learning at Killip Elementary.

THE FOUR CORE SYSTEMS

1. Assessment and Data

Assessment can be seen as tedious and time-consuming. Schools without a strong and valued assessment system tend to focus on hands-on work, such as engineering STEM activities, and stay away from a more balanced integration with the core English language arts (ELA) and mathematics standards. Then, when performance on state assessments drops, STEM becomes the culprit! Having a strong and valued assessment system that provides timely feedback on balanced integration can support your school district as the students experience change and learning begins to shift at your school. Once that transformation in learning becomes ingrained as practice, it's the assessment system that provides the continuity that can demonstrate success. In turn, those assessment results can become points of pride worthy of celebrating.

All Killip students are benchmarked and then progress-monitored four times throughout the year. Students needing additional support can be monitored as often as weekly, to collect data for informing changes to the interventions. Common grade-level assessments are developed for each STEM unit, and they include all content areas. Teachers also use myriad other formative assessments during instruction that provide information on student learning trajectories.

2. Common Planning Time for Teachers

Critical for the successful implementation of STEM education is the ability to provide common and collaborative planning time. This should be in place before you attempt to develop a schoolwide STEM curriculum. At Killip, the staff have rearranged their building schedule so that each grade level has all its students in special classes at the same time. This allows the teachers to meet once a week for an hour. During this planning time, the team conversations focus on the curriculum, instruction, and assessment. It is during this time that much of the curriculum development occurs. For longer or particularly complex units, such as the second-grade Pond Habitat unit (see Chapter 8), the school brought in substitutes so that the teachers could meet with an instructional coach and the STEM coordinator to plan the unit over the course of a full day.

3. Continuous Improvement Cycle

With its planning time systems in place, Killip has been able to develop a culture of continuous improvement in the school (see Figure 10.1). This provides a safety net that encourages open-mindedness, allows for failure, and rewards innovation. The units the teachers develop undergo an annual review and revision process in which the participating teachers reflect on the instructional goals for the unit and the effectiveness of meeting those goals and discuss possible innovations to improve the unit. The second-grade Pond Habitat STEM unit was not a real-world, problem-driven, fully integrated unit from the start—it was the result of five iterations using this continuous improvement process.

Figure 10.1. Killip Elementary's Continuous Improvement Model

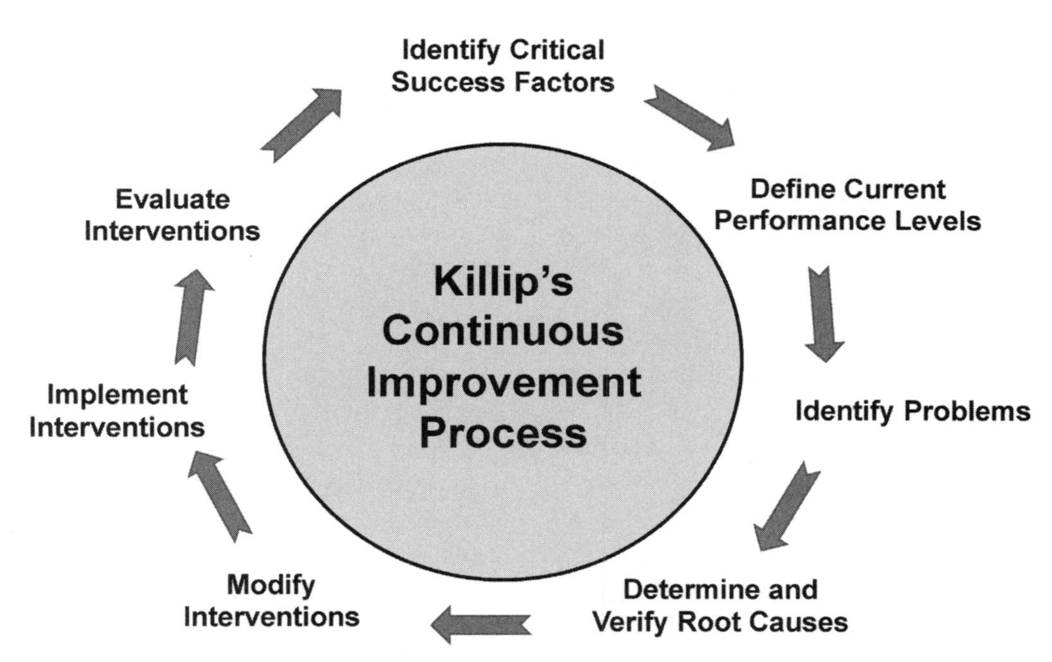

4. Incorporating Community Collaboration

Collaboration takes on many different meanings in the development of each STEM unit. All members of the school community take part in identifying ways to work together. It may be working with reading resource staff to secure age-appropriate texts, connecting with media specialists to locate digital or multimedia resources, or identifying community partners who work in STEM fields related to the concepts being considered. A part of the Killip team collaboration approach is community partner involvement, because not every teacher is going to be an expert across all content areas. That has become a big benefit of the collaborative approach that Killip has adopted.

When the second-grade team decided on the development of a STEM-focused pond habitat unit, they reached out to an Arizona Game and Fish ranger, an aquatic ecologist, and a landscaper who installs ponds for a local landscaping contractor. Althought teachers and schools can sometimes be hesitant about contacting community partners for their assistance, the Killip team found out that most experts in the community are happy to lend their support to the students. Schools should have a clear plan developed so that they can carefully explain the goal of their project and how the community partner may be able to help.

Moving From an Underperforming to a High-Achieving School

As described by Principal Joe Gutierrez

In 2005, Killip Elementary School was labeled as an underperforming school, as measured by the school accountability system of No Child Left Behind. Understanding that we needed to align our curricula, instruction, assessments, and interventions with the standards, we were able to raise our performance level to "Performing Plus" by 2008. As a school community, we pondered the question, "Is this enough for our students, so that they can move forward, obtain the necessary skills to become contributing members of society, and move themselves out of the vicious cycle of poverty"? The conclusion was no. The standards provide our students with an expected level of academic knowledge and skills, but what was missing was the application of this knowledge to the real world.

Poverty was first addressed with the enactment of the Elementary and Secondary Education Act (ESEA) in 1964 as part of President Lyndon B. Johnson's "War on Poverty," and it has been the most far-reaching federal legislation affecting education ever passed by the U.S. Congress. In 2002, Congress reauthorized ESEA, requiring states to administer assessments in reading and math to students in grades 3 through 12 to measure student achievement. States, school districts, schools, and students are now provided with a label depending on the student performance level as measured by these assessments. No Child Left Behind has implemented sanctions to serve as consequences for schools that earn low ratings. Yet, poverty still remains the number-one factor that contributes to the student achievement gap.

The Killip leadership team implemented a cycle of continuous improvement to evaluate the processes used to determine the effectiveness of our practices. What we found was that we needed to balance the knowledge and skills that we were teaching our students with a level of application that required them to apply those skills and knowledge in the real world—in other words, supporting our students to

Continued

(continued)

be successful as they matured in the 21st century. Through our research, we found that STEM had the potential to give us a framework to support our outcome of a balanced curriculum that would provide our students with skills, knowledge, and application. Research also told us that children of poverty lacked experiences and opportunities in the real world that would allow them to make relevant connections between what is learned in school and the purpose for that learning.

Killip was fortunate to learn that the Helios Foundation was awarding a competitive three-year grant to Arizona schools that had a desire to implement STEM in their school. Using the Arizona STEM Network's STEM Immersion Matrix, which was provided to us as part of the grant application, we found that Killip was at an Exploratory level of STEM implementation. We applied for the grant with the goal of moving the school from the Exploratory level to a level of Full Immersion as identified by the Arizona STEM Network's (2018) *STEM Immersion Guide.* We completed the grant application and submitted it, and after going through a rigorous review process, we were notified that we were one of seven schools to be awarded the grant!

To move from an Exploratory level to a Full Immersion level, we had to move our offering of STEM from being an extracurricular activity to one being included during our regular school day. As we identified our current practices, we found that if STEM were being included in our teaching practices at all, it was being taught in isolation and in a very limited amount of time. We concluded that if we were going to include STEM instruction during the school day, we would have to do it with an integrated approach.

Our team developed an action plan that required our STEM coordinator, our instructional specialists, and our grade-level teachers to create STEM units that would integrate the contents of the curriculum. The units were designed to be aligned with the Arizona College and Career Ready Standards and the *Next Generation Science Standards.* The units would include problem-based learning, would be student-centric, and needed to have some tangible solution or product at their conclusion. It was imperative that student activities require the use of 21st-century skills, that they be rigorous, that technology be used for instruction, and that students also use technology to reinforce their learning. A Full Immersion unit would include community and business outreach in which students would be supported in discovering work and career explorations.

It is important to realize that educational transformation of this magnitude requires a three- to five-year plan that includes activities that embed STEM into the culture and the climate of the school. There should be a realization that resources need to be provided to support professional development at all levels. It is critical to evaluate the plan, the framework,

Continued

(*continued*)

the process, the units, the instruction, and the students' achievement. Using a cycle of continuous improvement, we must monitor and adjust the initiative as needed based on the outcomes and the feedback of all team members.

If I had to describe all of these efforts in one word, I would say it would be *perseverance.*

What results has Killip Elementary School seen in the past five years? We have collected both quantitative and qualitative data to measure the implementation of STEM teaching. We have found that based on the AzMERIT state assessment in reading and ELA, the Killip students are growing and learning. The growth rate of Killip students moving out of Levels 1 and 2 to Levels 3 and 4 is the third highest in the district. Although the percentage of students earning a 3 or 4 is still not as high as the state or district level, we are making steady gains every year. As measured by the AzMERIT in science, which is only measured in fourth grade in the elementary schools, the percentage of students moving to higher achievement levels is increasing each year.

The level of students engaged in their learning is also increasing each year. The skill level of our students who were in kindergarten five years ago has increased, as measured by presentations made to the community when they seek monetary funding to support our projects. In addition, the number of fifth-grade students applying to STEM magnet programs in our middle school has increased by approximately 17%.

Student presentations using 21st- century skills have also improved each year. The students have increased their skills and have produced a higher quality of projects that have been funded from year to year. We now have a greater number of student leaders taking responsibility for schoolwide composting, working in our school garden, and taking care of Luna Park, which is a natural habitat for native plants, birds, and fish in our pond. All this is derived from the STEM units we created. And our students have been increasingly recognized for their work, receiving several awards at both the city level and the state level.

As measured by teacher evaluations, the use of technology has increased. Teachers say that their questioning strategies have become more rigorous and intentional. Student activities are moving from pencil and paper to more student-centered, hands-on demonstrations of understanding. Students' proficiency has improved as their learning has become more project and problem based, as opposed to relying on lower-level information or memorization tasks. The teachers are motivated and have been open to trying new things. A greater number of teachers are becoming teacher leaders in STEM.

Although we do not have quantitative data to support that our students will remove themselves from the vicious cycle of poverty, the experiences and

Continued

(*continued*)

opportunities that they have been exposed to and continuously participate in offer them a window into future possibilities. Along with the increased rigor and more personalized, relevant experiences tied to instruction, one can anticipate that the growth the students are making annually on the state-mandated assessments provides evidence that, by implementing STEM teaching and learning in our school, we have given our students a tangible reason to learn and the aspiration that they can achieve success as part of a future workforce.

CREATING A CULTURE FOR STEM IN THE K–2 CLASSROOM

Throughout this book, we have provided examples of K–2 STEM teaching and learning in several classrooms and offered suggestions on how these lessons and units could be implemented into your classroom setting or school.

We have seen how integrating the different STEM disciplines into a single, more cohesive learning experience is crafted over time with careful planning and reflection by the teachers. The rewards of this approach are many. As stated in a recent report, *Charting a Course for Success*,

> *The Nation is stronger when all Americans benefit from an education that provides a strong STEM foundation for fully engaging in and contributing to their communities, and for succeeding in STEM-related careers. … STEM education teaches thinking and problem-solving skills that are transferable to many other endeavors. STEM literacy gives individuals a better capacity to make informed choices on personal health and nutrition, entertainment, transportation, cybersecurity, financial management, and parenting. A STEM-literate public will be better equipped to conduct thoughtful analysis and to sort through problems, propose innovative solutions, and handle rapid technological change, and will be better prepared to participate in civil society as jurors, voters, and consumers.* (NSTC 2018, p. 5)

STEM teaching and learning provides students with the important reason to learn because students actually are applying what they are learning in new and meaningful ways. When disciplines converge in meaningful and engaging ways, rich learning experiences arise. As students focus on real-world problems, they begin to understand the complex issues and challenges inherent in finding solutions.

In *Charting a Course for Success*, the Executive Summary concludes,

> *The character of STEM education itself has been evolving from a set of overlapping disciplines into a more integrated and interdisciplinary approach to learning*

> *and skill development. … It seeks to impart skills such as critical thinking and problem solving along with soft skills such as cooperation and adaptability. Basic STEM concepts are best learned at an early age—in elementary and secondary school—because they are prerequisites to career technical training, to advanced college-level and graduate study, and to increasing one's technical skills in the workplace.* (NSTC 2018, p. v)

This is exactly what providing STEM experiences for students does. As we instill in our primary school learners the abilities to read and write, let's remember that STEM learning gives them something meaningful to read and something noteworthy to write about.

REFERENCES

Arizona STEM Network. 2018. STEM immersion guide. *https://stemguide.sfaz.org.*

National Research Council (NRC). 2014. *STEM integration in K–12 education: Status, prospects, and an agenda for research.* Washington, DC: National Academies Press.

National Science and Technology Council (NSTC). 2018. *Charting a course for success: America's strategy for STEM education.* Washington, DC: NSTC, Committee on STEM Education.

Weld, J. 2017. *Creating a STEM culture for teaching and learning.* Arlington, VA: NSTA Press.

BIBLIOGRAPHY

Aguirre-Muñoz, Z., and M. L. Pantoya. 2016. Engineering literacy and engagement in kindergarten classrooms. *Journal of Engineering Education* 105 (4): 630–654.

Bybee, R. W. 2011. Scientific and engineering practices in K–12 classrooms: Understanding *A Framework for K–12 Science Education. Science Scope* 35 (4): 6–11.

Early Childhood STEM Working Group. 2017. *Early STEM matters: Providing high-quality STEM experiences for all young learners.* Policy report, Erikson Institute and University of Chicago STEM Education, January. *http://ecstem.uchicago.edu.*

Hernandez, P. R., P. W. Schultz, M. Estrada, A. Woodcock, and R. C. Chance. 2013. Sustaining optimal motivation: A longitudinal analysis of interventions to broaden participation of underrepresented students in STEM. *Journal of Educational Psychology* 105 (1): 89–107. *www.ncbi.nlm.nih.gov/pmc/articles/PMC3838411/pdf/nihms512414.pdf.*

Lister, J. 2017. Crossing boundaries: The future of science education. *Scientific American*, August 15. *https://blogs.scientificamerican.com/observations/crossing-boundaries-the-future-of-science-education/?print=true.*

McClure, E. R., L. Guernsey, D. H. Clements, S. N. Bales, J. Nichols, N. Kendall-Taylor, and M. H. Levine. 2017. *STEM starts early: Grounding science, technology, engineering, and math education in early childhood.* New York: Joan Ganz Cooney Center at Sesame Workshop.

Weld, J. 2018. *Charting a course for success: America's strategy for STEM education.* Washington, DC: Committee on STEM Education of the National Science and Technology Council.

Note: Page references in **bold face** indicate information contained in figures or tables.

INDEX

INDEX

INDEX